BEEKEEPING STUDY NO'
for the
BBKA MICROSCOPY EXAMI'

Prepared by:

J.D.YATES B.Sc.(Hons), C.Eng., FIEE
and
B.D.YATES SRN, SCM

BEE BOOKS NEW AND OLD
Ash View, Tump Lane,
Much Birch,
Hereford,
Herefordshire HR2 8GP

First published 1995
Second Edition 2004
Third Edition 2014

ISBN 978-0-905652-37-5

Printed and bound in Great Britain by
Print Plus, 126, Widemarsh Street,
Hereford HR4 9HN

PREFACE

The slow but gradual demise of the County Bee Lecturers is now complete due to financial cut backs by county council educational authorities. The old CBLs used to do invaluable work in the diagnosis of adult bee diseases in addition to providing a host of other services to beekeepers. The National Bee Unit (NBU) used to provide a diagnostic service for adult bee diseases free of charge but, because of Government policy whereby they have to be self financing, they now levy a charge plus VAT for this service. Because the charge is so high, beekeepers are reluctant to use the service.

As a result of these developments most County Associations of the BBKA are now organising and training interested beekeepers to do their own adult bee diagnostic work with the objective that eventually each branch will be self sufficient with their own competent microscopists. It was with this in mind that the BBKA Examinations Board decided to instigate a Microscopy Certificate to provide a measure of competence for those interested in this aspect of our craft.

There is no one set book where all the detailed information for the syllabus (which is the table of contents in this volume) can be found. The preparation of these notes is an attempt to fill this need at an economical price and without recourse to many library visits.

The Microscopy Certificate was instituted in 1994 and the syllabus has undergone various changes by the Examination Board over the years. Some of the changes, in our opinion, have been good but others are considered to have been unnecessary tinkering. The EU have had their sticky fingers in the pie too, with health and safety requirements now tacked on the end of the syllabus at section 8. We are indebted to John Chandler of the Harrogate & Ripon BKA for the information that he has kindly provided as an Appendix for this new section.

This is the third edition of these notes and it is by popular request that they have been prepared. The aim of the Examination Board is to maintain a similar standard between different Examiners in different parts of the country. Writing the notes concentrates the mind and makes one think carefully about the depth of knowledge required. When the Board approved the syllabus for publication it will be clear that a certain standard was envisaged and Examiners will be required to examine candidates to this standard. It is hoped that the notes will not only give guidance to Candidates but to future Examiners too about the depth of knowledge required from the Candidate. The notes should also prove useful for those contemplating the NDB practical examination.

JD & BD Yates, Newton Ferrers, March 2014.

The BBKA state:

Aims.

This Examination aims to provide a qualification and measure of achievement for those beekeepers with an interest in pollen, anatomy, pests and diseases relevant to the honeybee. A certificate holder should be able to give beekeepers practical guidance on the selection and use of microscopes, help them identify a range of pests and bee diseases from samples provided and give them advice on where to find further sources of information. A pass in the Microscopy Certificate gives exemption to the Microscopy Section of the Advanced Husbandry Assessment.

Conditions of Entry.

The Candidate shall have passed the BBKA Basic Examination or an equivalent Examination approved by the Board. The date on which the certificate was obtained shall be entered on the application Form.

The Assessment.

(a) Examiners approved by the Board shall conduct the Examination. The Board may wish a trainee Examiner or a member of the Board to be present as an observer, but prior approval of the Candidate must be obtained.
(b) The Candidate shall provide:
Two suitable microscopes, one for dissection and one for the examination of microscope slides (a compound microscope with x400 magnification and incorporating an oil immersion lens)
Dissecting tools and instruments
Spare slides and cover slips
6 pollen slides and 4 anatomy slides made by the candidate from the lists provided in the syllabus
Approximately 40 freshly killed worker bees.
The equipment required to embed one or more bees in wax for dissection purposes
Any other equipment the candidate may require.
(c) A requirement of the Examination is the assessment of the material prepared by the Candidate.
(d) The Examination shall be conducted at a place and time convenient to both the Examiner and the Candidate and shall be of an oral and practical nature.
(e) The Examination may be expected to be completed within 2½ to 3 hours.

CONTENTS

1.0 CONSTRUCTION AND PARTS OF A MICROSCOPE

The Candidate shall discuss with the Examiner:

1.1 The essential differences between microscopes used for dissection and those used for examining the detail on smears and specimens down to about 0.25μm in size.
1.2 The difference between reflected light and transmitted light for illuminating the object and how these are achieved in the construction of a microscope.
1.3 The concept of lens magnification for both a single lens and a compound system of lenses in simple terms only.
1.4 The purpose of the principal parts of the dissecting microscope.
1.5 The purpose of the principal parts of the high power microscope.

2.0 PRINCIPLES AND THEORY OF THE LIGHT MICROSCOPES

The Candidate shall discuss with the Examiner:

2.1 The range of magnification required for a dissecting microscope suitable for dissecting a honeybee and how this range is achieved.
2.2 The range of magnification of a compound microscope suitable for examining specimens for the detection of honeybee diseases except those caused by viruses.
2.3 The minimum sized object that can be seen using a light microscope and an elementary understanding of the dependence of this on the wavelength of light and the numerical aperture of the objective.
2.4 The functions of the stage, condenser/mirror, diaphragm, eyepiece, objective lenses, coarse and fine focus, in the high power microscope.
2.5 The optical features to be taken into consideration in the choice of a microscope. For example, good resolution, minimal distortion of image, a flat optical field, par focal and spring-loaded objectives.
2.6 What is meant by the term 'depth of field' and its importance.
2.7 The use of oil immersion for higher magnifications and the significance of the refractive index of the oil.
2.8 The advantages of using filters of different colours.
2.9 The use of an eyepiece graticule and its calibration.

3.0 SETTING UP MICROSCOPES FOR BEE DISEASES AND POLLEN IDENTIFICATION

The Candidate shall demonstrate to the Examiner:

3.1 The setting up of a dissecting microscope for the identification of Acarine.
3.2 The setting up of a high power microscope for the identification of Nosema and Amoeba.
3.3 The setting up of a high power microscope for oil immersion to view very small specimens eg pollen and bacteria.

The Candidate shall discuss with the Examiner:

3.4 The magnification required for the identification of Acarine, Nosema, Amoeba, AFB and EFB giving the approximate size of the pathogens.
3.5 The magnification required for the identification of pollen giving the approximate range of sizes of pollen grains commonly collected by the honeybee in the UK.

4.0 DIAGNOSIS OF ADULT BEE DISEASES AND COLONY INFESTATIONS

The Candidate shall demonstrate to the Examiner:

4.1 The dissection and examination of a worker bee for the presence of Acarine.
4.2 The preparation and examination of a sample of bees for Nosema and Amoeba,

The Candidate shall discuss with the Examiner:

4.3 The identification of worker bees with signs of Deformed wing virus or Chronic paralysis virus from specimens or images provided by the Examiner.
4.4 The identification of two pests from specimens, slides or images provided by the examiner and discuss the anatomical features that enabled this identification. These pests will be selected by the Examiner from the following list:

Varroa destructor, Tropilaelaps sp, Braula coeca and the larval, pupal and adult stages of small hive beetle, greater and lesser wax moths.

4.5 The size of the sample required for the examination of adult bee diseases and its statistical significance.
4.6 How and where the adult bee sample should be taken from the hive and the reasons involved.
4.7 The assessment of the level of infection or infestation and likely outcomes if treatment is withheld.
4.8 What advice should be given to beekeepers on the actions to be taken and sources of information in the event of an adult bee disease or colony infestation being identified.

5.0 DIAGNOSIS OF BROOD DISEASES

The Candidate shall discuss with the Examiner:

5.1 The key features of American and European Foul brood including the difference between healthy and diseased larvae.
5.2 What advice should be given to beekeepers on actions to be taken/ sources of information in the event of AFB /EFB being identified.
5.3 The identification of chalk brood from specimens or images provided by the Examiner.

6.0 POLLEN IDENTIFICATION

The Candidate shall discuss with the Examiner:

6.1 The general construction of a pollen grain.
6.2 The collection and preparation of pollen from (a) flowers, (b) pollen loads from the honeybee, (c) honey.
6.3 Six slides, made by the candidate, labelled with the date the slide was made, the scientific name and the approximate size, selected from the following list of pollen grains:
forget-me-not, dandelion, rape, lime, sycamore, poached egg plant, crocus, willow, heather, hogweed, rosemary, hawthorn, hazel.
6.4 How the slides were made and how they should be stored for long term use.
6.5 How the size of the pollen on the slides was determined.
6.6 Three pollen slides provided by the examiner.
6.7 An outline of how microscopic analysis can be used to determine the floral sources and geographic origin of honey samples including the need to take into account the over and under representation of pollen in a multi-floral honey.
6.8 How the presence of honeydew in a honey sample can be detected by microscopic examination.

7. DISSECTION AND ANATOMY OF THE HONEYBEE

7.1 The Candidate shall provide freshly killed workers, demonstrate ability to embed them in wax during the examination and be able to perform and discuss the abdominal dissection as requested by the Examiner.
7.2 The Candidate shall make and provide for discussion four labelled anatomical slides with one slide made from each of the following four lists. At least three of these slides should be prepared as permanent hard mounts .

List A	List B	List C	List D
Front, middle and hind leg of worker	Mouthparts of worker (displayed)	Fore and hind wings of worker or drone	Sperm
Comparative slide of a hind leg from drone queen & worker	Comparative slide of mandibles of drone, queen and worker	Antennae of both worker and drone	Two Varroa (one mounted dorsally and the other ventrally
Everted endothallus of drone genitalia	Sting of a worker	Portion of trachea	Hypopharyngeal gland

8.0 HEALTH AND SAFETY

The Candidate shall discuss with the Examiner:

8.1 The potential hazards of working with chemicals, naked flames, microscopes, electrical devices & dissecting instruments.

8.2 The need to undertake a risk assessment before commencing any activity of a practical nature.
8.3 The need to wear protective clothing, chemical resistant gloves and goggles when handling hazardous chemicals and to always work safely with respect to themselves and others.
8.4 The safe disposal of hazardous waste e.g chemicals, broken glass slides, scalpel blades and remains of bees.

APPENDICES

INDEX

DIAGRAMS

1. CONSTRUCTION AND PARTS OF A MICROSCOPE

Discuss with the Examiner:

1.1 The essential differences in microscopes used for dissection and those used for examining the detail on smears and specimens down to about 0.25µm in size.

The essential differences are easily summarised in tabular form as shown below:

ITEM	DISSECTING	COMPOUND	REMARKS
IMAGE	Erect	Inverted	Erect for dissecting purposes
MAGNIFICATION	Low (c. x 40 max.)	High (c. x 1200 max.)	Magnification changed by a combination of objectives and eyepieces in both types
LIGHTING	Normally reflected	Normally transmitted	Light source may be internal or external in both cases
EYEPIECES	Binocular prismatic	Binocular or monocular	Binocular is essential for serious dissecting work
COST	Cheap c., £120	Expensive c., £200	Suitable for this examination

In the case of the compound microscope, it is only being used for observation and measurement so it matters little that the image is upside down. However, in the case of the dissecting microscope it is essential that the image is 'the right way up'. It would be impossible to work if the object was moved to the right and its image moved to the left. It should be noted that both types of microscope are technically compound, however, the term ‘compound’ has been in use for many, many years referring to the higher magnification instrument. Therefore, we do not propose deviating from the generally accepted norm.

1.2 The difference between reflected light and transmitted light for illuminating the object and how these are achieved in the construction of a microscope.

1.2.1 Reflected light.

In order to see the image in either type of microscope it has to be illuminated and this can be achieved by shining light onto an object and observing the light that is reflected from that object. Our normal vision, of the world around us, operates on this principle and it is also generally employed in most dissecting microscopes. There is an interaction between the object, the illumination and the eye. For example, it is possible to see a speck of dust in a sunbeam which is only possible because the particle scatters the sunlight falling on it and reflects it to our eye. Under normal circumstances it is highly unlikely that the minute particle would be visible without recourse to some magnification. Here it might be a good point to note a few definitions as follows:

Opaque - not transmitting light, not transparent
Transparent - transmitting rays of light without diffusion
Translucent - transmitting light but not transparent

In general, there are no special setting up techniques required for using reflected light with a dissecting microscope. Sufficient light intensity is required in order to illuminate the work being undertaken. The illumination is said to be **incident** illumination and the angle of incidence can vary from about 90^0 to a few degrees. External light sources are generally used. Fibre optic lights are now quite common enabling light to be accurately directed where required.

1.2.2 Transmitted light.

Transmitted light is used to observe say a colour slide by illuminating it from behind and observing the light that is transmitted through the actual slide. Similarly, when we look at a glass of wine, the light must be behind it. The same principle is used in the compound microscope which requires a source of illumination on the opposite side of the slide to that being observed. Clearly, the slide or object must be translucent, thereby allowing light to pass through enabling it to be observed. Absorption takes place in the specimen and thickness is characterised by brightness variations.

It will be clear that an optical system is required to provide the transmitted light used in the compound microscope and this optical system is required to interface with the optical system used to magnify the object (ie. the contents of the slide). The interface point is in the object being observed and the two systems must be critically focussed at the interface point. Once this is understood the setting up becomes much easier to achieve in practice.

1.3 The concept of lens magnification for both a single lens and a compound system of lenses in simple terms only.

If light is passed through a prism it will undergo two refractions as shown in diagram 1 below. If prisms of differing sizes are constructed as shown, and dimensioned so that they all come to a focus at the same point, it will be clear that as the number increases they will eventually approach the smooth curve of a lens. The refractions all focussing at one discrete point is only true if the light is monochromatic (of one frequency only). If no corrections are made to the lens, the lens is said to suffer **chromatic aberration**. Lens manufacturers make corrections to the lenses (particularly the objectives and eyepieces) in order that they focus correctly for all colours, from blue through to red, at the same point. Unfortunately, correction is never completely perfect and emphasises the use of single colour filters. The corrected lens is said to be **achromatic**. It should be noted that dispersion into the various colours takes place at both refractions in the prism; for simplicity the dispersion is only shown at the last refraction in the diagram.

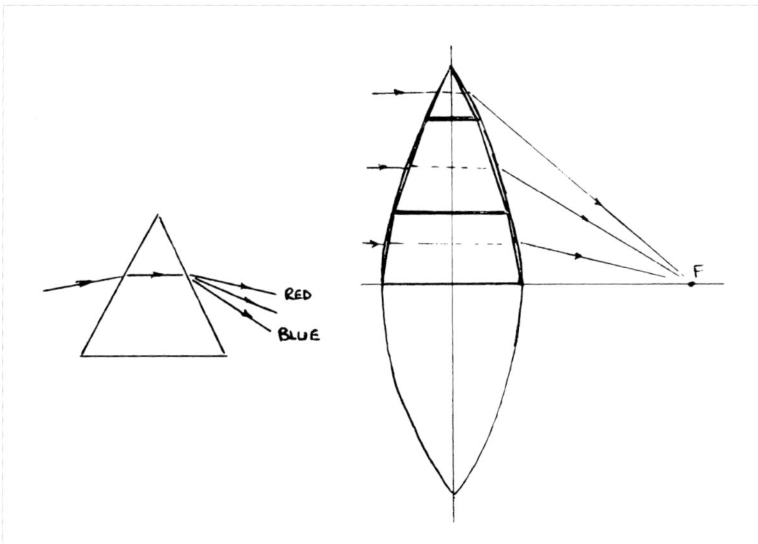

Diagram 1. Light passing through prisms, the basis of a spherical lens.

The magnification of a single lens is shown below:

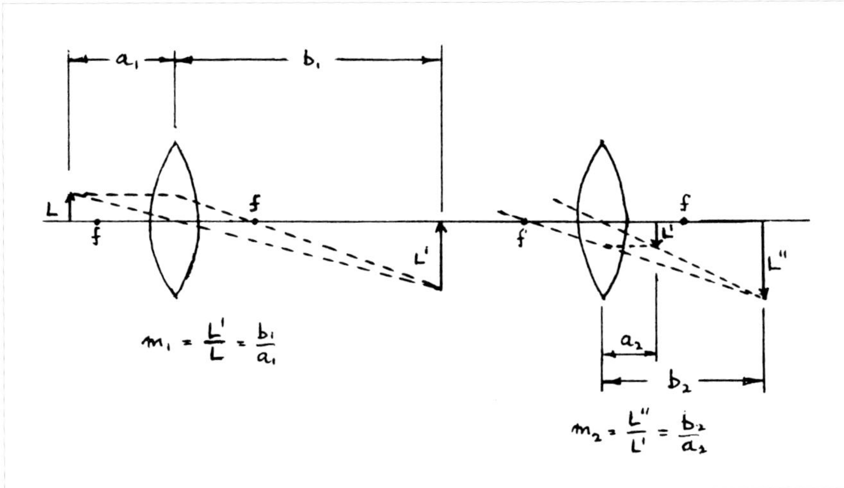

Diagram 2. The magnification of a single lens.

There are two cases, one when the object is at a distance from the lens greater than its focal length and two, when it is at a distance less than the focal length. In the first case the image is real (it can be displayed on a screen) and in the second case the image is virtual (it cannot be displayed on a screen). The second case is, of course, the simple magnifying glass.

The derivation of the formulae for the magnification is simply just a case of comparing similar triangles. Diagram 3 shows the optics of the compound microscope which combines the two single lens systems discussed above.

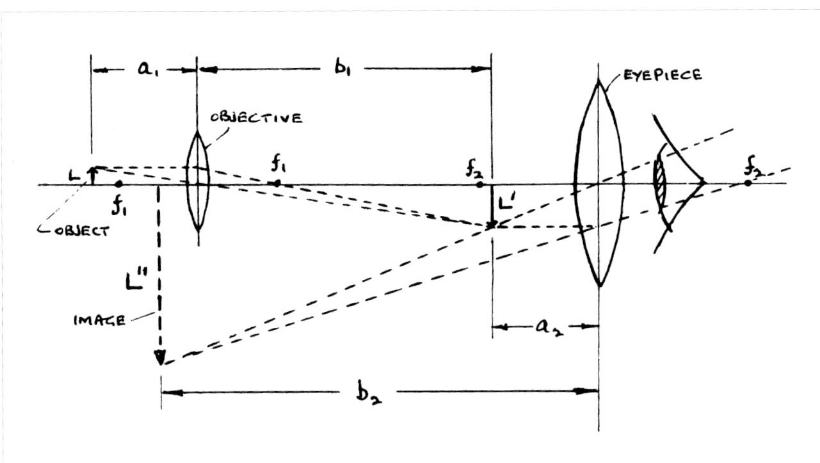

Diagram 3.The optics of a compound microscope.

It will be seen that the magnification of the compound microscope is as follows:

L = size of the object L' = size of image of lens f'_1 = object of lens f_2
and L" = final image and the image of lens f_2
By similar triangles $L \div a_1 = L' \div b_1$ and $L' \div a_2 = L'' \div b_2$
Magnification (M) = $L'' \div L = b_1 b_2 \div a_1 a_2$
Which by similar reasoning is the magnification of lens f_1 (m_1) x the magnification of lens f_2 (m_2), which is $m_1 = L' \div L = b_1 \div a_1$ and $m_2 = b_2 \div a_2$
Also it will be seen that $M = m_1 \times m_2$ which is the power of the objective multiplied by the power of the eyepiece quoted in section 2.2.2.

1.4 The purpose of the principal parts of the dissecting microscope.

Reference should be made to diagram 4 below. The principal parts of a dissecting microscope, sometimes called a stereo microscope, are similar to the compound microscope but without the sub-stage and the stage and are as follows:

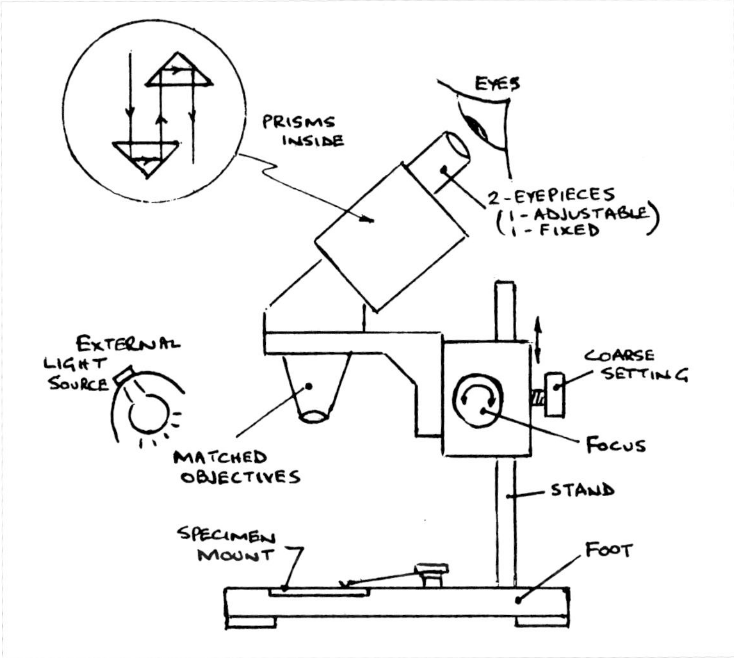

Diagram 4. A typical dissecting microscope with external light source.

Note: Some microscopes can have an internal light source.

Foot - the metallic base of the instrument which should be heavy in order to bring the overall centre of gravity low and have a wide base for stability purposes. It usually incorporates arrangements for holding the specimen or object. The more expensive models incorporate a built in light source for transmitted light.
Limb - very often fixed to the base or the foot.
Binocular system - attached to the limb (or stand) with provision for varying the distance between the objective lenses and the object on the foot. Coarse and fine focussing is essential. As stated previously, one of the essential features of the dissecting microscope is that the image should be erect and movement of the object will create a movement of the image in the same direction (see diagram 5 below). This is achieved by incorporating two right angled isosceles prisms (at right angles to each other) whereby light from the objective lens is refracted and reflected twice in each prism before being directed to the eyepiece. Many models have the coarse focussing by raising and lowering the whole binocular system up and down the limb and fixed with a locking screw. An important feature of this type of microscope is the distance between the objectives and the specimen as large as possible to give adequate room to work on a dissection.
Binocular eyepieces - one is fixed and one can be focussed to accommodate differences in the users' eyes. Most models have fixed magnification c. x 10.
Objective lenses - these are usually interchangeable say x2 and x4. With the x10 eyepieces this would give x20 and x40 capability.

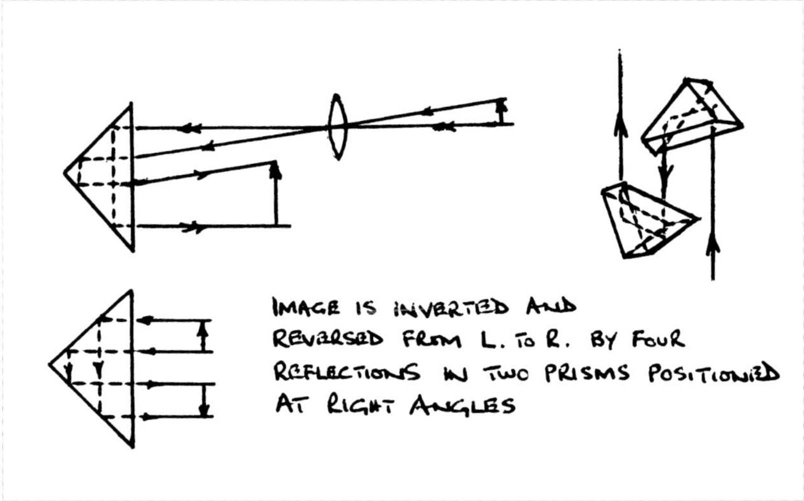

Diagram 5. Erecting prisms in a dissecting microscope.

1.5 The purpose of the principal parts of the high power microscope.

It should be noted that the Examination Board have incorrectly changed the wording of compound microscope into high powered microscope - we would regard an electron microscope as being high powered.

Reference should be made to diagram 6 showing a typical compound microscope with external light. The principal parts are listed also below with an explanation against each:

Foot - the metallic base of the instrument which should be heavy in order to bring the overall centre of gravity low and have a wide base for stability purposes.
Limb - this is attached to the foot, often on a swivel arrangement so that the angle can be adjusted for ease of use. The swivel bearing should have no slack, and should be provided with a satisfactory locking arrangement when set. The limb must be mechanically rigid because of the two optical systems which are mounted on it. As the instrument has to measure down to a fraction of a micron, no torsion or bending is permissible. Provision is made to vary the distance between the objective lens and the stage for focussing purposes. Coarse and fine focussing screws are essential.
Tube - this is mounted on the upper part of the limb and contains the eyepiece at the upper end and the objective at the lower end. It comprises the first discrete optical system. The eyepiece is often interchangeable and would have a value of say x10 combined with a turret of objective lenses say x4, x10 and x40 giving a capability of x40 to x400 magnification. The objective lens is to provide the basic magnification and resolution and the eyepiece is to provide the balance of the magnification and a suitable mount for any eyepiece graticule that may be used.
Stage - this is the part which supports the object (or slide) and is located near the swivel between the tube and the condenser / mirror. Again it should be of rigid construction.
Sub-stage assembly - this is located on the lower part of the limb and houses the condenser and / or mirror, a focussing screw for the condenser, an adjustable diaphragm to control the transmitted light, a filter holder and perhaps a built in light source. It comprises the second discrete optical system. The purpose of the condenser / mirror is to focus the light source in the object while the diaphragm controls the area of illumination and adjustment for minimising glare.

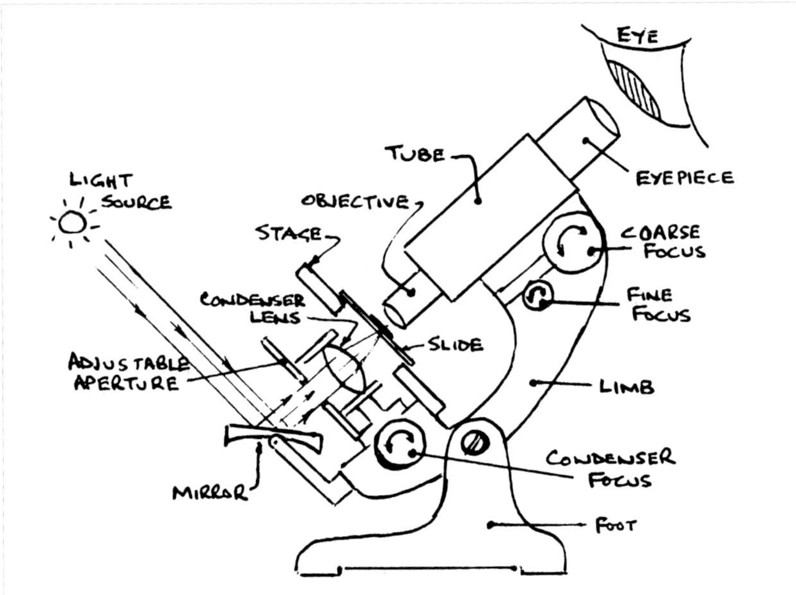

Diagram 6. A typical compound microscope with external light source.

All compound microscopes will have the tube, eyepiece and objective lenses. The simplest will have only a mirror below the stage which is normally two sided, one side flat and one side concave. More complicated models will have mirror and condenser and a built in light source. A most desirable optional extra must be a **mechanical stage** for making fine and controlled movements of the object slide. Any type of compound microscope suitable for the practical aspects of this examination may be used but you will be expected to know about the more complex models.

* ** *** ** *

2. PRINCIPLES AND THEORY OF THE LIGHT MICROSCOPES

Discuss with the Examiner:

2.1 The range of magnification required for a dissecting microscope suitable for dissecting a honeybee and how this range is achieved.

The range of magnification required for a dissecting microscope is very much determined by the operator and his / her dexterity working with dissecting tools under magnification. Generally, the ranges available are from about x5 to x40. Our own preference is working with a power of x20. A good compromise would be 15.

For those new to the world of microscope magnification, the total magnification (M) is the magnification of the eyepiece lens (m_e) multiplied by the magnification of the object lens (m_o). This is true for both types of microscopes. $M = m_e \times m_o$.

Thus, if the object under examination was, for example, 0.25mm diameter the image would appear to be 0.25mm x 20 = 5.00mm diameter if the total magnification of the instrument was x20. Similarly, an object 1mm long would produce an image 20mm long.

2.2 The range of magnification of a compound microscope suitable for examining specimens for the detection of honeybee diseases except those caused by viruses.

2.2.1 The pathogens and their sizes.

Firstly, consider the measurements used;

1 metre			= 1m
1m	÷	1000	= 1 mm (milli-metre)
1mm	÷	1000	= 1µm (micro-metre or micron)
1µm	÷	1000	= 1 nm (nano-metre)

A virus is the smallest pathogen. It cannot be seen under a light microscope as it is too small. The range of these pathogens is from c. 6nm up to c. 400nm and includes, among many others, sacbrood virus, chronic bee paralysis virus, cloudy wing virus, etc. An electron microscope is required to see an image of a virus and it was not until after the last war (1939 - 45) that they became available using the technology developed during the conflict.

Bacteria, fungi and protozoa are the next largest group of pathogens and range in size from c. 0.1µm to 20µm but the ones that we are interested in start at c.1µm and include those causing the following bee diseases:

Nosema and Malpighamoeba,
AFB and EFB, together with
Chalk brood and Stone brood.

2.2.2 Range of magnification required.

It will be clear that the objects required to be examined range from c. 1µm to 20µm and this is readily achieved with the compound microscope which can have a range of magnification from c. x40 to c. x1200.

It is necessary to examine the amount required and this depends on the size of the image. The human eye can readily resolve lines in a graticule one hundred to the inch (0.01in) and under ideal circumstances four hundred to the inch (0.0025in) at a distance of about 9in which is known as the **'near point' or the 'distance of distinct vision',** the nearest point that an object can be brought to the eye to have 'a closer look'. This latter resolution is equivalent to an angle equal to one minute of arc (60 minutes in 1 degree) and assumes normal vision. To achieve such performance requires optimum lighting of the object so this is a necessary factor which must be taken into consideration. A 'comfortable' sized image for most people would be 1mm and on this basis the following magnifications would be required:

object = 20µm - magnification = x100 - image = 2000µm = 2mm
nosema spore = 6µm - magnification = x400 - image = 2.4mm
AFB spore = 1µm - magnification = x1000 - image = 1mm

Therefore, with a compound microscope with an eyepiece of x10 and a range of objective lenses on the turret of x10, x40 and x100 diagnosis of all the bee diseases except those caused by a virus can be confirmed. On our own compound microscope we have a x4 objective giving an overall power of x40 which we find very useful for examining, for example, nymph and adult *Acarapis woodi* in dissected trachea.
The optics of the compound microscope are arranged so that the final image is always at the 'near point' thus making the physical size of the instrument compact. In order to do this it is necessary for the object lens to have a very short focal length with its image falling within the focal length of the eyepiece as shown previously in diagram 3.

The focal length of the objective must be about 0.25in for an objective of about x50. If the focal length was say about 1in the length of the microscope would have to be about 50in which would be a practical impossibility. Additionally, it is essential that as much light as possible from each point of the object should enter the objective and pass through the microscope otherwise the greatly magnified image will be too faint to be seen distinctly. This means that the diameter of the objective must subtend as large an angle as possible at the object which involves making the objective with the largest relative aperture possible.

The calculation of the overall power of the compound microscope is extremely simple and is shown in section 1.3. From diagram 3:

$M = m_1 \times m_2$ or the magnification of the objective times that of the eyepiece.

A further point of interest should be noted in connection with the short focal length of the objectives. The very short focal lengths brings them in very close proximity to the object particularly at the higher powers so great care should be taken when adjusting the microscope so that the objective does not come into physical contact with the object. When objectives of x100 are used the object and the objective are connected together via a special oil [see section 2.7].

2.3 The minimum sized object that can be seen using a light microscope and an elementary understanding of the dependence of this on the wavelength of light and the numerical aperture of the objective.

The wavelength of light.

The minimum sized object that can be seen using a light microscope depends on the wavelength (λ) of the light being used.

It is common knowledge that light is an electro-magnetic wave motion similar to radio waves but of a very much shorter wavelength and higher frequency. The relationship between frequency (f) and wavelength (λ) is as follows:

wavelength (λ)	x	frequency (f)	=	constant
metres (m)	x	MHz	=	300

The electro-magnetic spectrum is roughly as follows, in ascending order of frequency:
Long waves (1500m), medium waves (650m - 200m), short waves (150m - 10m), VHF (3m - 1.5m), UHF (1m - 0.5m), microwaves (150cm - 15cm), X-rays (mm), infra red (below 800nm), white light (800nm - 400nm) and UV (400nm - 300nm)

If 'white light' is passed through a prism and focussed on a flat surface a continuous spectrum is visible (to man) which extends from red to blue where $\lambda_{red} = 800$ nm and $\lambda_{blue} = 400$ nm. This is due to the different frequencies being refracted at different angles, blue suffering the greatest deviation. Most people have done this during their school days. If we change nm into μm, then $\lambda_{red} = 0.8$ μm and $\lambda_{blue} = 0.4$μm. These wavelengths are smaller than our 1μm spores for AFB, a size that is approaching the resolving limit of a light microscope. For a variety of complicated reasons, which are not required for the syllabus, **the absolute resolving power is half the wavelength of the light ($\lambda/2$) being used**. Thus if blue light was being used, $\lambda/2_{blue} = 0.2$μm and a realistic minimum size that could be resolved as a discernible image would be 0.25μm. The minimum size would double to 0.5μm if red light was being used.

Thus if the compound microscope is being used for examining the detail of objects below say 3μm in size, it is prudent to use a blue filter between the light source and the object. In fact it is always advisable to use a single coloured source (monochromatic) to ensure the best definition of the image. A variety of effects take place including refraction and diffraction which causes aberration of the image. Aberration due to refraction problems is reduced by using monochromatic light hence the reason for using filters.

Numerical aperture of the objective.

The numerical aperture of an objective lens = NA = $\mu \sin\theta$

where μ = refractive index in the medium containing the object or specimen
and θ = axial angle made by the most oblique ray from the object to the image.

As shown in diagram 7 below, NA corresponds with the critical angle in refraction according to Snell's Law, where the critical angle r corresponds to the angle θ.

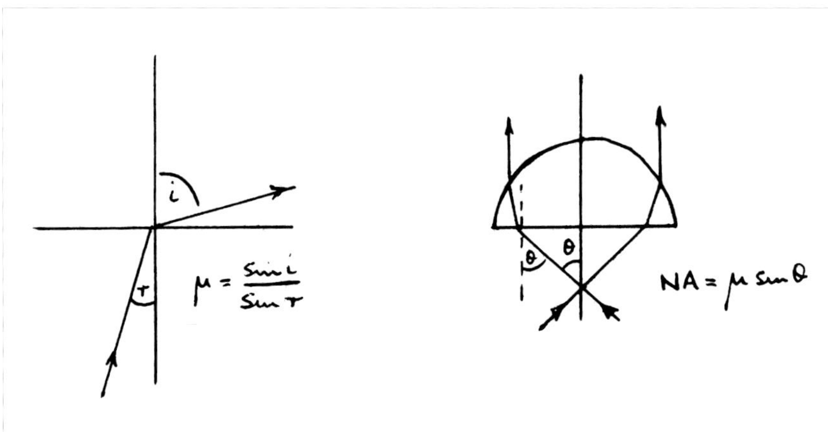

Diagram 7. Relationship between Snell's Law and NA

The angle θ is governed by the actual mechanical contours of the lens and the refractive index m. This imposes theoretical limits for the NA of air, water and oil immersion objectives of approximately 1.00, 1.33 and 1.52. The practical limit for these values are 0.95, 1.25 and 1.4 respectively giving an axial angle of about 72^0. It will be clear that the higher the NA of a lens the better its light gathering power (larger axial angle) and the greater its resolving power which is directly proportional to its NA, illustrated in diagram 8. In this diagram both lenses have the same NA and hence the same axial angle at the object.

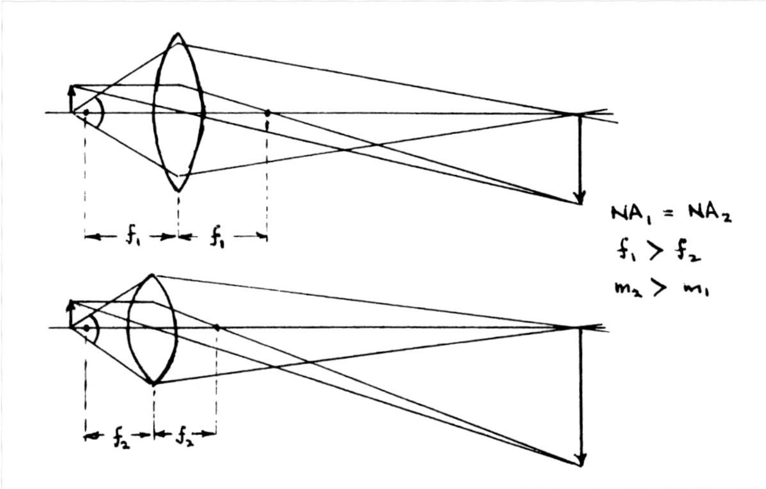

Diagram 8. Resolving power is proportional to NA

The NA of a lens, particularly the objective of a microscope, is of the utmost importance because the resolving power depends directly on this quantity. The relationship between NA and resolving power is a linear one and an objective with NA = 1.2 will resolve twice as much detail as an objective with an NA = 0.6. See diagram 8 above

It is to be noted that a magnification value of 1000 x NA is often quoted as being the approximate limit beyond which further magnification will produce no more detail. This readily gives a method of calculating the maximum magnification which can be used in the eyepiece as shown in the table below.

Magnification (objective) m_1	NA	1000 x NA = total M	Eyepiece magnification $m_2 = (1000 \times NA) \div m_1$
x 4	0.1	100	x 25
x10	0.25	250	x 25
x 40	0.65	650	x 16
x 100	1.2	1,200	x 12

This theoretically calculated eyepiece magnification takes no account of the light proceeding up the microscope tube and into the pupil of the observer's eye. The light rays appear to emerge from the eyepiece in a small disc of rays called the **exit pupil** or **Ramsden Circle (RC)** the diameter of which is given by the formula:

RC(dia) = (2 x distance of least vision x NA) ÷ M = (500 x NA) ÷ M.

However, if RC (dia) is fixed at 1.5mm (size of a normal pupil of the eye), then

M = (500 x NA) ÷ 1.5 = 333 x NA where the distance of least vision = 250mm

Thus the table above modifies to:

m_1	NA	333 x NA = total M	Calculated m_2 $= M \div m_1$
4	0.1	33	8
10	0.25	83.25	8
40	0.65	216.5	5
100	1.2	399.6	4

In practice the nearest preferred size eyepiece would be selected. Our own microscope has the first 3 objectives and NAs but the microscope was only supplied with one eyepiece (x 10).

It will now be abundantly clear that NA is an important parameter which must be born in mind when undertaking any practical microscopy work.

2.4 The functions of the stage, condenser/mirror, diaphragm, eyepiece, objective lenses, coarse and fine focus, in the high power microscope.

The objective and the eyepiece, as we have already seen, is one of the two discrete optical systems making up the compound microscope. The other is the illumination system controlled by the condenser and or the mirror. The two systems interface at or in the object on top of the slide and under the cover slip. The interfacing is extremely critical and if it is not correct it will not be possible to resolve detail down to about 1 μm.

2.4.1 The stage.

This is the part of the microscope that supports the object (usually between the slide and the cover slip) and is located between the two optical systems, that is, the illuminating system below and the magnification system above. It requires to be constructed rigidly and often provides fixing holes for the attachment of a mechanical stage. Mechanical stages are relatively expensive but virtually essential for any serious measurement work. Some compound microscopes can be found whereby the stage and sub-stage can be adjusted with respect to the fixed tube for focussing purposes.

2.4.2 The condenser / mirror.

The function of this part of the instrument is twofold, one, to provide satisfactory illumination of the object or specimen and two, to produce a cone of light to exploit the full power of the objective. It is the part of the compound microscope which seems to cause most difficulty particularly when setting up the instrument for use. If the principle of the optics is understood then the setting up becomes a relatively simple matter. The sub-stage settings have the capability of forming the final image and its quality. In the more expensive microscopes provision is found for centring the condenser with the objective lenses being used; note that the syllabus does not require a Candidate to set up the centring of a compound microscope. The basis of the condenser / mirror optics is to collect the rays of light from the illumination source (lamp or blue sky) and to use the condenser lens to focus these at a point in the object being examined or observed. The rays of illuminating light will then be diverging after passing through the objective lens. The condenser is a simple single lens which generally is optically inferior to those used for the objectives and eyepiece. If an inbuilt light source is used then the mirror becomes redundant and is usually removed from the instrument.

The condenser controls the quality of the final image, its visibility and its resolution. It enhances the illumination of the object to provide an image bright enough when enlarged and to do this it provides a cone of light from the object into the objective to use the full power of this lens. If the condenser is not correctly set up the final image will be marred. It cannot be overemphasised the importance of the condenser adjustment for final image quality. Ideally the condenser ought to be corrected to a similar standard to that of the objective. In most instruments in the cheaper price range this is never the case and for the purposes of this examination such microscopes are unnecessary. The condenser should produce an even illumination over the entire field and to achieve this, the most important correction is for spherical aberration.

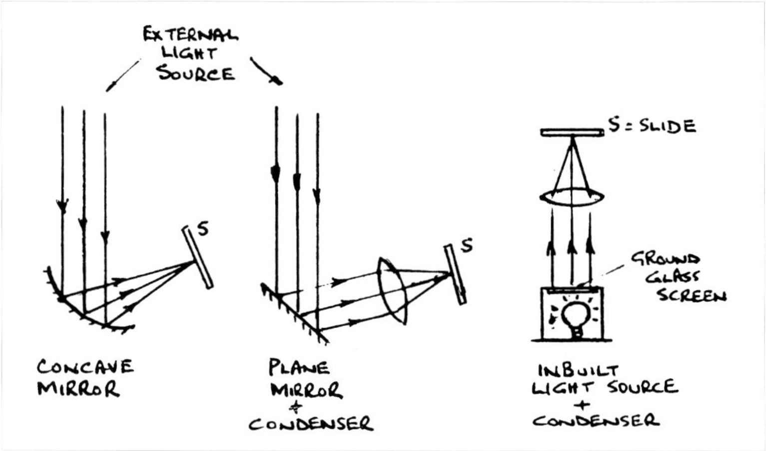

Diagram 9. Possible condenser/mirror configuration.

When there is no inbuilt light source and an external one is used the mirror (the plane side) acts only as a reflector to direct the light in the correct direction into the condenser lens. On cheaper instruments where no condenser lens is provided the mirror is used in conjunction with an external light source but in this case the concave side is used which focuses the parallel beam of light directly into the objective. Most instruments of this type have a fixed position for the mirror and the manufacturer ensures that the focal length of the concave mirror is correct to provide the illumination required. Diagram 9 shows the arrangements possible.

The final point about the sub-stage assembly is the provision of a holder to accommodate a filter to provide a single coloured source and a variable aperture, both are found on most instruments in the medium price range.

2.4.3 Diaphragm.

The purpose of the diaphragm, sometimes called a field or lamp diaphragm, is to vary the effective diameter of the illuminating source by varying the effective diameter of the condenser. The aperture is changed by varying the diaphragm diameter (opening and closing). It is an essential component in the adjustment of the illuminating system and should be opened to a maximum size whereby the image is not deteriorating due to haze or glare. It is often recommended that the condenser aperture should be about 75% of the objective aperture for best results which was initially postulated by the microscopist E.M.Nelson. We prefer to adjust the diaphragm for the best image.

2.4.4 The eyepiece.

The function of the eyepiece is (a) to provide the balance of the magnification, (b) to produce the final **virtual image** at the near point and (c) to provide a mount for any graticule and pointers that may be fitted. The rays of light from the virtual image are converging and thus can be arranged to enter the pupil of the eye. The instrument must be designed with the real image of the objective falling within the focal length of the eyepiece and at the point where provision has been made for the graticule and pointer to be mounted. That is, these become the object for the eyepiece and the final image is made up of an image of the original object with an image of the graticule and pointer superimposed upon it, all three being in focus at the near point.

2.4.5 The objective.

We have already seen that in order to design an instrument with a tube of usable length it is necessary to have a lens with a very short focal length with the widest aperture to gather as much light from the object as possible. The greatest magnification of a lens of a given diameter is when it approaches a sphere; in order to increase the magnification further it is necessary to reduce the diameter. Such lenses have a small radius of curvature and hence a high magnification but because of the reduced diameter limit the amount of light that can pass through them. Therefore, in general, the function of the objective is to provide the major part of the magnification required and to produce an image inside the focal length of the eyepiece so that the eyepiece acts as a magnifying glass. The objective is the most important of the lens systems and provides all the detail to be observed. The image which it produces is a **real image** which could only be observed on a plain screen, rather like a cinamatograph, because the rays of light are diverging.

It is essential that the objective is mounted in such a way to ensure that its axis coincides with the axis of the tube (ie. the objective is centred correctly). When the objective lenses are turret mounted it is important that the turret is robust with no slack in the mechanism to ensure precise centring when changing from one lens to another. The image from each lens does not necessarily fall in exactly the same position in the eyepiece and it is necessary to re-focus when changing objectives. In a high class instrument refocussing is not required when objectives are changed by rotation of the turret and the instrument is then said to be **par focal**. If the microscope is being used continually it makes for easier operation but for the average beekeeper this is not an essential feature. Our own compound microscope is not par focal but by the judicious use of washers of varying thickness used on the screw threads of the objective lenses for fixing to the turret it is now virtually par focal requiring a minimum of re-focussing. A further refinement in the more expensive instruments is the incorporation of spring loaded objective lenses to prevent damage to the lens when working at high powers with the lens very close to the slide.

2.4.6 Coarse and fine focus.

The purpose of the coarse and fine focussing screws is really obvious and need little explanation. Clearly both are used for focussing the microscope and the coarse one is always used first to obtain the best image and then it is 'fine focussed' with the fine control.

On an average compound microscope the coarse control provides about 20mm of movement for one turn of the coarse control whereas one turn on the fine control produces 0.2mm of movement. The gearing ratio is, therefore, 100:1.

2.5 The optical features to be taken into consideration in the choice of a microscope. For example, good resolution, minimal distortion of image, a flat optical field, par focal and spring-loaded objectives.

2.5.1 Resolution

There are three factors involved in discerning the detail in an image; these are the object itself, light and the eye with its light receptors on the retina where the image is formed. In section 2.2.2 the resolution of a normal eye was discussed which can resolve 1 minute of arc under good lighting conditions. The limit of resolution is when the details in the image are too close to fall on separate retinal receptors and 1 minute of arc is approaching this limit when viewed from about 10in, the distance of least vision or the near point. Diagram 10 below shows the eye and how the image is enlarged by having a closer look.

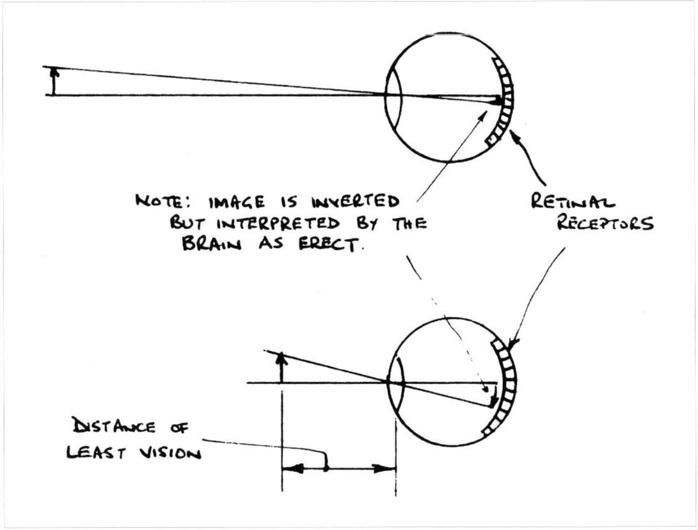

Diagram 10. The resolution of the eye.

As soon as recourse is made to optical instruments to magnify the object, light will be lost, so the illumination of the object becomes critical at high magnifications as shown in diagram 11 below.

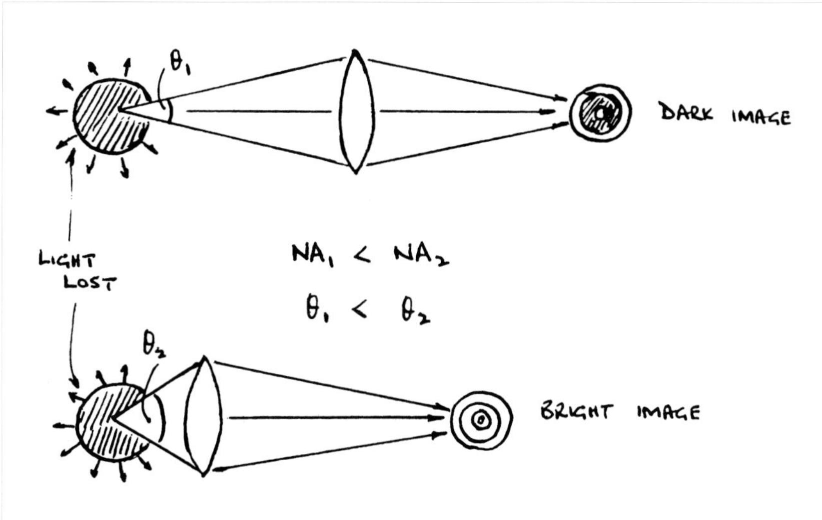

Diagram 11. Light from the object forming the image.

Diagram 8 shows that the resolving power is proportional to the 'light grasp' of the objective lens which we have seen corresponds to the numerical aperture. A distinction must now be made between the resolving power (of a microscope) and resolution (of the eye). Diagram 12 shows the process of events producing an image of a very small circle containing a pattern of extremely fine dots.

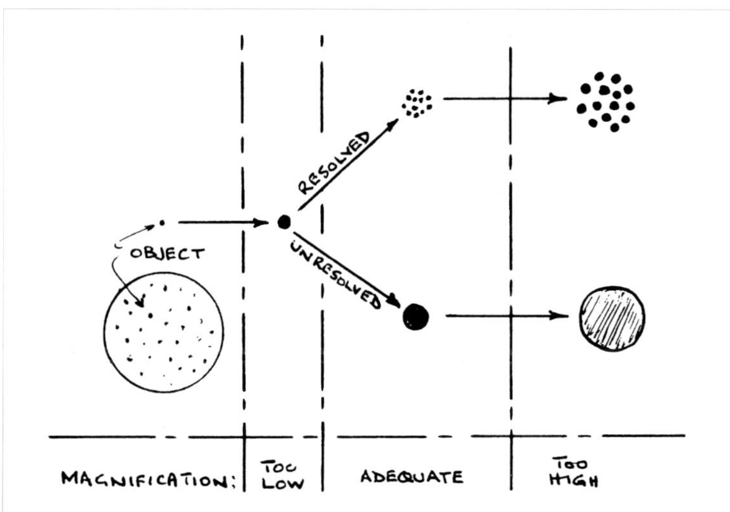

Diagram 12. Magnification and resolution.

It will be clear from the above diagram that providing the objective lens has an adequate NA then the detail in the image can be resolved. However, if the NA is inadequate the detail cannot be resolved. Further magnification of either of these two conditions achieves nothing.

Finally, we can say that the microscope is an instrument for rendering fine detail visible but in order to do so the microscopist must form an image whereby the detail can be seen. This involves setting up the instrument correctly in respect of the lighting source and the resolving power of the instrument.

2.5.2 Minimal distortion of the image.

There are three major defects of the lenses in a microscope which cause degrading of the image, namely, spherical aberration, coma and astigmatism which affect the definition of the image while the fourth defect, distortion affects the position of the image.

It is beyond the scope of these notes to explain these defects in detail but the manufacturer always corrects the lenses for all of them. However, the amount of correction is reflected in the price of the instrument. Both spherical aberration and coma can be reduced by decreasing the radius of the stop. Spherical aberration is in very simple terms the condition whereby the marginal rays of a pencil of rays come to focus at different distances from the defined focus of a lens. Coma is serious for points close to the axis of a lens due to the asymmetrical position of the object relative to the lens. Astigmatism is a condition whereby the image is displaced, being most serious for points at an appreciable distance from the axis of the lens. Distortion occurs when a lens produces an image of an object at a position closer to the axis than that formed along the auxiliary axis. It increases as the stop is moved further from the lens and for the same reason is zero when the stop coincides with the lens.

When choosing a microscope it is always best to try it with a couple of known slides to ensure that it comes up to expectation.

2.5.3 A flat optical field.

A further refinement found in more expensive instruments is flat optical field eyepieces, a correction that enables the whole field of view to be in focus at one setting. In cheaper microscopes without this facility the edges of the field are out of focus when the centre of the field is in focus.

2.5.4 Par focal and spring loaded objectives.

Par focussing and spring loaded objectives were discussed in section 2.4.5.

2.6 What is meant by the term 'depth of field' and its importance.

The depth of field is some times referred to as the 'depth of focus' which to our mind is a more realistic description of the phenomenon. The eye cannot distinguish between a point and a circle of 0.1mm diameter, an important concept when considering the focussed image. The best that a normal eye can resolve is 1minute of arc at a distance of 250mm = 0.073mm, approaching the 0.1mm diameter circle mentioned earlier.

Now if a lens produces a point image of a point object then it will produce a circular image of another point at a distance 't' from the original point object. The diameter (d) of the circular image must be 0.1mm in order for the eye to be able to resolve it. The depth of field of the lens can then be shown to be = 2t:

$$2t = \{a^2d/(bD-ad)\} + \{a^2d/(bD+ad)\} \qquad \text{where}$$

a = distance of the object from the lens, b = distance of the image from the lens and D = diameter of the lens. It will be clear from the above formula that the depth of field increases with decreased D. D is, of course, controlled by the diaphragm and the smaller D becomes, the greater the depth of field. This is another consideration to be born in mind when adjusting the diaphragm in the setting up procedure. As 'a' decreases with increasing power of an objective lens, it will also be seen from the formula that the depth of field will also decrease and the lower the power, the greater the depth of field.

The depth of field is important when studying, for example, pollen grains in order to determine the optical section. An image will be produced when focussing on the pole and the image of the equatorial view will be blurred because of the short depth of field of a high powered lens. Generally, the depth of field of microscope objectives are very small, measured in microns, and quite unlike the camera where the depth of field can be measured in metres.

Microscopists should get used to preparing drawings of the microscopical appearance of each plane in focus through thick objects eg. pollen grains. It is not old fashioned to do so and it is an acquired art looking down the microscope with one eye while drawing and using the other eye to view the paper adjacent to the microscope all in one operation. Try it and the depth of field concept will never cause any further difficulty.

2.7 The use of oil immersion for higher magnifications and the significance of the refractive index of the oil.

If oil immersion is not used at high magnifications two problems will arise namely:

a) Spherical aberration due to refraction.
b) Loss of light from the object due to reflection.

2.7.1 Spherical aberration due to refraction.

At high magnification, that is with objectives say greater than x100, blurred images will be produced due to refraction effects when the rays of light leave the cover slip. This gives rise to multiple images being formed as shown in diagram 13 below.

When the surface of the cover slip and the objective are connected with a drop of oil that has the same refractive index as glass (n = 1.5) no refraction takes place due to the oil between the objective and the object. Thus the elimination of the air eliminates the initial spherical aberration due to refraction.

Note that spherical aberration can be axial or lateral.

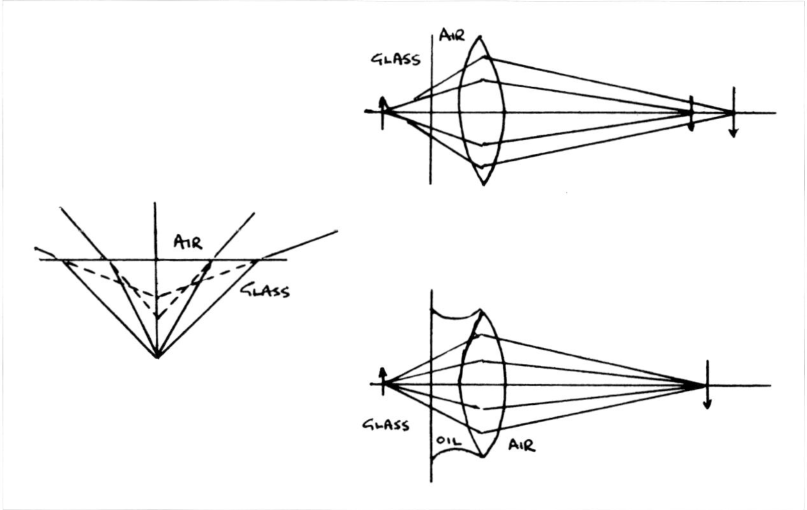

Diagram 13. Refraction effects with and without oil.

2.7.2 Loss of light due to reflection.

In diagram 14 below the rays of light are shown leaving the illuminated object in the glass slide both without and with oil eliminating the air gap.

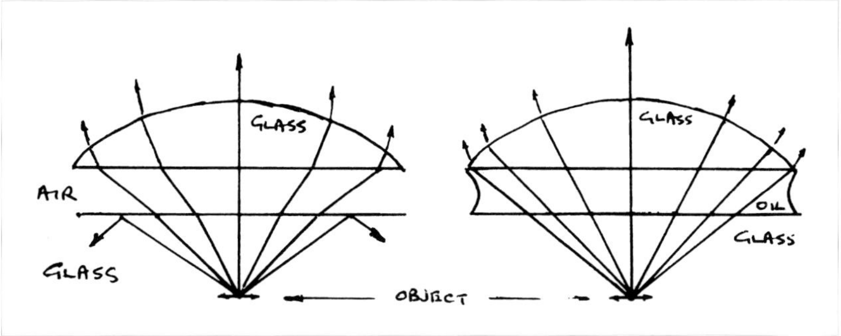

Diagram 14. Loss of illumination of the object due to reflection.

In the case without oil it will be clear that all rays greater than the critical angle will not leave the slide but will be reflected back and lost. Any ray where sin r > 1/n will always be reflected where n = refractive index of the glass (1.5). This means that when $r > 41^0$ all rays from the object are reflected.

2.7.3 Other points about oil immersion.

Oil immersion is sometimes referred to as 'homogeneous immersion'. There are some practical points relating to the application of immersion oil which should be observed as follows:

a) The objective lens is always very close to the object (slide cover slip) and great care is needed so that mechanical damage does not occur to either.
b) As it is not possible to alternate from medium power to high power it is essential that the microscope is well set up before the oil is applied particularly with respect to the illumination.
c) Oil immersion will destroy the specimen unless it is permanently mounted or it is a dried specimen (as for diagnosing AFB and EFB)
d) Immersion oil used to be oil of cedar wood with its refractive index adjusted to that of glass by adding Canada balsam. This mixture had the disadvantage of drying hard on the lens if it was exposed to atmosphere for too long making the objective lens difficult to clean.
e) Nowadays, synthetically prepared oils are used which do not cause this problem and it is stated that objectives are now specially designed to work with these oils.
f) Again it is stated that an oil of the wrong quality can dissolve the cement used in fixing the objective lens to its mount. We consider this to be somewhat remote with the use of modern epoxy resins.
g) It is more likely that the special liquids used to clean the oil from the lenses would be a more effective solvent.
h) If the oil is left on the lens it can damage the lens by the process of oxidisation acting on the glass.
i) When the lens is cleaned it should be done in three stages:
    1. Wipe the oil off with a lens paper.
    2. Clean again with a lens paper and a small amount of solvent (eg. xylol).
    3. Wipe with a dry lens paper.

2.8 The advantages of using filters of different colours.

There are 3 basic advantages in the use of single colour filters as follows:

1. The single coloured filter is a quick, inexpensive and easy way of providing something similar to a monochromatic source and hence it improves the definition of the image where lenses have not been fully corrected for chromatic aberration. This is useful where stain colours are not critical in the work being undertaken.

2. To enhance visibility or to reduce excessive contrast with stained specimens using filters of a complimentary or the same colour. eg. if the nucleus of a cell is stained blue and the surrounding cytoplasm is stained red then a red filter will give a black nucleus.

3. To enhance resolution by shortening the wavelength. Blue light (λ = 400nm) has twice the resolving power of red light (λ = 800nm); see section 2.3. UV would be even better if we could see it like the bee; never-the-less UV light is used for special work in photomicrography and the sensitive coating on the film responds to the UV light instead of the human eye.

2.9 The use of an eyepiece graticule and its calibration.

The eyepiece graticule (sometimes called an eyepiece micrometer) is a circular clear plastic disc with a small scale engraved on it from 0 to 100 divided into 10 main divisions each of which are sub-divided into 10. It looks like a 10cm ruler with mm divisions but of course much smaller. The engraving is very fine and 0 to 100 occupies a length of about 1cm. We regard one as essential for pollen identification work and they can be purchased very cheaply from the suppliers. For example, many of the rosaceae pollens are so similar in appearance that the only way of identification is by measurement. The present cost of a micrometer graticule is c. £15.

There are basically two ways of calibrating the graticule for our purposes which are as follows:

1. Comparison with a stage micrometer which is a slide with a specially engraved calibrated scale with a length of about 1mm divided into tenths and hundredths. The scale is produced by either the use of an engraving engine or by photographic reproduction and protected with a cover slip. The centre of the graticule image is adjusted to coincide with the centre of the stage micrometer and calibrated in both directions from the centre to the left and to the right. This method is the most expensive but more accurate than the second method. The present cost of a 10µm stage micrometer is c.£35.

2. By the use of *Corylus avellana* (hazel) pollen which is very constant in size (c. 25µm in diameter). A slide is made with the pollen and the graticule calibrated for each sized objective. The method is not as reliable as the micrometer slide. It should be remembered that the final size of a mounted pollen grain depends upon the mountant which is being used; for example, Sawyer postulates 25µm for the size of

hazel pollen while Kirk postulates 27μm, a difference of 2μm, ie. c. 7% difference. Because pollen grains swell due to preparation and the mountant used, it will be clear that this method is not as accurate as method 1 above.

Other methods for linear measurement include the use of various types of micrometer eyepieces whereby a cross wire traverses the field of view.

It should be noted that optical thickness can be measured by movement of the focal plane but only lateral linear measurements are required for this syllabus and examination.

* ** *** ** *

3. THE SETTING UP OF MICROSCOPES FOR BEE DISEASES AND POLLEN IDENTIFICATION

Demonstrate to the Examiner:

3.1 The setting up of a dissecting microscope for the identification of Acarine.

There is usually one adjustable eye piece (the other is fixed). Focus the microscope, using the focussing screw, on an object using one eye and the fixed eye piece. Then observe the object with the other eye through the adjustable eye piece. Focus again but use only the adjustment on the adjustable eye piece. Then use both eyes and both eye pieces by first adjusting the width between them for your own eyes to obtain the brightest circular field of view. Finally re-focus with the focussing screw to obtain the sharpest image.

An 'angle-poise' lamp or similar is required to illuminate the object and the position of this should be adjusted to obtain the best illumination for the job in hand noting that reflected light from the object will be observed. The low wattage long life bulbs are ideal giving bright light with minimum heat.

3.2 The setting up of a high power microscope for the identification of Nosema and Amoeba.

The setting up process is not simple and it is well to understand that there are two optical systems involved. The first is the microscope itself involving the compound eye piece and the objective lens. This system starts at the object being observed (ie. a minute portion of the slide) and finishes with the image on the retina of the eye. The second is the illumination system which starts at the illumination source, the sky or an electric lamp, and finishes as a focussed image in the slide and illuminating a very small portion of it. The two systems have to be aligned very accurately with both coinciding at the same point in the slide. In the best microscopes there is provision for adjusting and centring the light source with the objective lens system in a horizontal plane and, once set, seldom requires adjustment. The Examiners would never require the 'centring' to be set or adjusted for this examination.

It is setting up the illumination system that seems to give most Candidates trouble and we believe this is because they do not have a mental picture of the associated optics [see section 2]. Consider a magnifying glass and burning a hole in a bit of paper by using the sun - everybody has done this in their youth. In essence, the same principle is used in order to set up the illumination system, but NOT using the sun. A bright blue sky is ideal. The illumination system collects parallel rays of light and brings them to a focus in the slide. The system will have either a simple concave mirror which will do the focussing or an adjustable condenser lens (similar to the magnifying lens burning the paper) together with a plane mirror. The mirror, used with the condenser lens, does no focussing but simply changes the direction of the light beam by reflection. Don't try the concave mirror with the condenser - it won't focus. If the microscope has a built in light source it is unlikely to have a mirror. The only other bit of equipment is a variable aperture (constructed rather like a camera shutter) associated with the condenser lens and a filter holder.

The variable parts of the magnification system are the coarse and fine focussing screws and usually a selection of three objective lenses mounted on a rotating turret. A typical microscope would have an eye piece x10 and 3 objective lenses x4, x10 and x40 giving magnifications of 40, 100 and 400 respectively. An optional extra for the compound microscope is a mechanical stage to hold a standard slide which can then be moved in two directions, normal (at right angles) to each other, by means of threaded thumb screws. These stages are well worth while but are not cheap and not essential for diagnosing nosema and amoeba. One other optional extra, which is very cheap and should be fitted, is a graticule which fits in the eyepiece. It can be calibrated for all the magnifications used on the microscope thereby allowing any image to be measured very accurately.

Now we are ready to start setting up the microscope for the Examiner. Let us assume that it has a built in light, condenser lens, aperture control and has a total magnifying capability of x40, x100 and x400 as above. Proceed as follows:

1. Switch on the light source and put in, for example, a prepared nosema slide or any other suitable slide for focussing purposes only.
2. Select the lowest magnification (x40) and using the coarse screw, focus for the sharpest image. Open the aperture if there isn't enough light and then finally focus with the fine screw for the sharpest image (the use of the fine focus screw may be found to be unnecessary at this low magnification).
3. Remove the slide and do not touch the fine or coarse focussing screws.
4. Now focus the condenser by producing the sharpest image of the ground glass covering the actual lamp (this cover is the light source as far as the microscope optics are concerned). It should have a granular appearance and there may be some coloration of the image due to a cheap form of condenser lens which matters little for our present purpose. Adjust the aperture to give a comfortable light intensity while focussing.
5. Now replace the slide and re-focus with the fine screw. Reduce the aperture to the minimum light intensity for viewing the slide and then re-focus again using the fine control. The microscope is now set up with critical illumination for x40 magnification and for the particular thickness of slide that is being used. If the slide thickness is changed (and slides are available in differing thicknesses) the microscope will require to be readjusted completely.
6. Change the turret to provide a magnification of x100. If it is a good microscope it will not require refocussing, it is said to be par focal, ie. the objective lenses focus in the same focal plane. Most of us cannot afford this luxury and we are involved in a bit of extra work of refocussing each time the magnification is changed.
7. Repeat operations 1 to 5 inclusive and the microscope should be set up for x100 magnification.
8. Set the turret for x400 total magnification and repeat 1 to 5 in order to set up the microscope for maximum magnification. Take great care not to drive the object lens into the glass slide, at x400 magnification and above, the end of the lens will be very close to the glass slide because as the magnification increases the focal length decreases. It should now be possible to seek out those nosema spores, they should show up like white rice grains (the illumination intensity will determine how white) and you should know roughly what size they should appear.

One further refinement which is worth mentioning is that some microscopes have provision for a filter between the light and the condenser. The filters can be a variety of colours or even ground glass to act as a diffuser. By using a blue filter it will generally provide a better image as it is limiting the colour spectrum and cutting out the orange and red wavelengths. Theoretically, a lens only focusses at a given point at a particular wavelength and correcting it to focus at the same point for all visible wavelengths is a costly procedure which is reflected in the price of the instrument.

The different configurations available on compound microscopes together with combinations of various light sources are:

a) no mirror, condenser, inbuilt lamp (as discussed above),
b) mirror, condenser, external lamp (eg. 'angle-poise') or blue sky,
c) mirror, no condenser, external lamp or natural light from the sky.

Setting the microscope up for b) is identical to a) except that the plane mirror is used to reflect the bulb of the 'angle-poise' lamp into the condenser. Initially, adjust the mirror for maximum light in the field of view and focus the condenser on the surface of the bulb (use the lettering '240v 60w', it is easy to see) as in 4 above (or put a mark on the bulb with a felt tipped pen). If a bright blue sky is used, put a mark (a bit of drafting tape) on the window pane and focus on that in order to get critical illumination.

Setting the microscope up for c) use the concave side of the mirror and adjust initially for maximum light intensity. The focal length of the concave mirror should match the microscope but you will now have realised that this is the cheap option and there is no adjustment. Such a microscope has limitations which are related to the focussing of a concave mirror and are beyond the scope of this syllabus.

A final word concerning the setting up the condenser aperture; some books state that the last operation in the setting up procedure is to remove the eyepiece and adjust the condenser aperture until about 66% of the back lens of the objective is illuminated. We do not use this method ourselves as we like to keep the set screw retaining the eyepiece in place in the tube; many find the method confusing. The method we have described is, in our opinion, easier for the casual user of compound microscopes.

The nosema spore has a bipolar filament coiled round twin nuclei. At x400 magnification, and if the microscope is set up correctly, it is just possible to distinguish the two black nuclei. Additionally, if you have a good AFB or EFB slide it is just possible to distinguish the individual spores or bacteria at x400 magnification. Both these are simple tests to provide a measure of your prowess in setting up and if you can do this on a strange microscope you are unlikely to have much trouble in the examination on this part of the syllabus.

3.3 The setting up of a high power microscope for oil immersion to view very small specimens, eg. pollen and bacteria.

Most pollen, even the smallest, can be observed without resorting to oil immersion which is usually required for the bacteria causing AFB and EFB. The spores of AFB and EFB are often described as being similar to nosema (ie. rice grained in shape but considerably smaller) or diamond shaped with all four corners rounded off. On the major axis both types of spore are between 0.75μm and 1.0μm long. If a magnification of x400 is used, a 1μm spore would have an image size of 400μm = 0.4mm and an image of 1mm if a magnification of x1000 is used. The latter is the normal magnification for AFB and EFB diagnosis and an oil immersion objective will be required.

Generally, a dry slide is used, that is one that has the smear dried out without a cover slip. The microscope is set up in exactly the same way as previously using a slide without a cover slip. Ensure that, with the x100 objective, very great care is taken to focus on the top surface of the slide without causing mechanical damage. There is often sufficient dust on an old slide for this purpose; it is best to view the microscope from the side and bring the objective down until it is just not touching the slide and then raise the objective with the fine focussing screw. Familiarise yourself with the direction of rotation of the controls which are anticlockwise to raise the objective lens and vice versa.

Having set the microscope up in the dry position the actual slide is inserted and a small droplet of immersion oil is put on the slide and the microscope lowered until it makes contact with the oil. Again view from the side while doing this and then finally focus. It will be clear from section 2 that the image should appear brighter and it may be necessary to reduce the illumination. Do not forget to clean the objective when work is completed as outlined previously.

3.4 The magnification required for the identification of Acarine, Nosema, Amoeba, AFB, and EFB giving the approximate size of pathogens.

This part of the syllabus should now be very simple and the required magnifications are given in the table below for the disease pathogens.

Pathogen	Size	Magnification	Image size
Acarapis woodi	c.120 x 65µm	x40	4.8 x 2.6mm
Nosema apis	5 to 8µm	x400	2 to 3.2mm
Malpighamoeba mellificae	10 to 12µm	x400	4 to 4.8mm
Ascosphaera apis	Spores c. 2µm Spore cysts c. 12µm Fruiting bodies c. 60µm	x400 x400 x400	0.8mm 4.8mm 24mm
AFB & EFB	0.75 to 1.0µm	1000	0.75 to 1.0mm

3.5 The magnification required for the identification of pollen giving the approximate range of sizes of pollen grains commonly collected by the honeybee in the UK.

The size of pollen grains range from c. 5µm (Forget-me-not, *Myosotis spp.*) to c. 127µm (Evening primrose, *Oenothera biennis*) with the majority falling within 15 to 50µm range. It will be clear that a magnification of x400 would be quite adequate giving image sizes from 2 mm to 50.8mm.

* ** *** ** *

4. DIAGNOSIS OF THE ADULT BEE DISEASES

Demonstrate to the Examiner:

4.1 The dissection and examination of a worker bee for the presence of Acarine.

Diagnosis of acarine can only be made with certainty by examination of the 1st thoracic trachea, the only place that *Acarapis woodi* Rennie will infest. It should be noted that there are other types of acarapis, which infest external parts of the bee, such as the wing bases, that do not form part of the syllabus. A simple dissection is required to expose the 1st thoracic trachea, the apparatus required to do this is listed below:

a) A dissecting microscope providing an erect image with a magnification of between 10 and 20 and a small spot light or screened reading lamp.
b) A pair of forceps with extremely fine points honed with a carborundum stone and checked by viewing under the microscope. Hone and check, hone and check again until the points are perfect giving a good grip at the extreme points. A tweak with a pair of long nosed pliers may be required if the points are not meeting squarely at their ends.
c) Acarine needles are simply made with #8 stainless steel sewing needles pushed eye end first into a small piece of soft wood with a pair of long nosed pliers. It is important that the prongs, mounted in the wood, are parallel and not greater than $^1/_{16}$in apart. When they are absolutely correct a dab of 'Araldite' (quick setting epoxy resin) should be put at the point where the needles enter the wood thus bonding the wood and the needles together. We use 3 acarine needles to examine three bees at a time.
d) A dissecting needle with handle. Again, make your own with a #8 needle, it will be better and cheaper than one you can buy. Make a second one with a finer needle.
e) A sloping surface to mount the bee on with the acarine needle. Cork is often recommended but we find high density white polystyrene foam is best which is used widely as a packing material. A 2in cube cut in half along a diagonal with a junior hack saw will provide a pair of suitable blocks. In our own examinations of samples we use a long sloping block whereby we can mount three bees at a time.
f) A pair of forceps (large) with rounded tips for handling individual bees and holding them in place while fixing them to the block with the acarine needle.

Dissection.

It goes without saying that the bees have to be dead before they are dissected and the fresher the bees the better. After any animal dies autolysis (decomposition) starts immediately and the higher the temperature the faster this process occurs (the bacteria multiply more quickly). Thus it is always desirable to kill the sample by freezing and keep it at a low temperature until it is ready to be examined. The bees must be pliable for dissection and will have to thaw out if they have just been removed from the deep freeze. A judicious warm up in the microwave oven can complete this process if time is at a premium but don't ruin your sample by inexperienced cooking.

The first job to do is to mount the bees on the block and remove the first pair of legs and the head complete in one operation as shown in diagram 15 below.

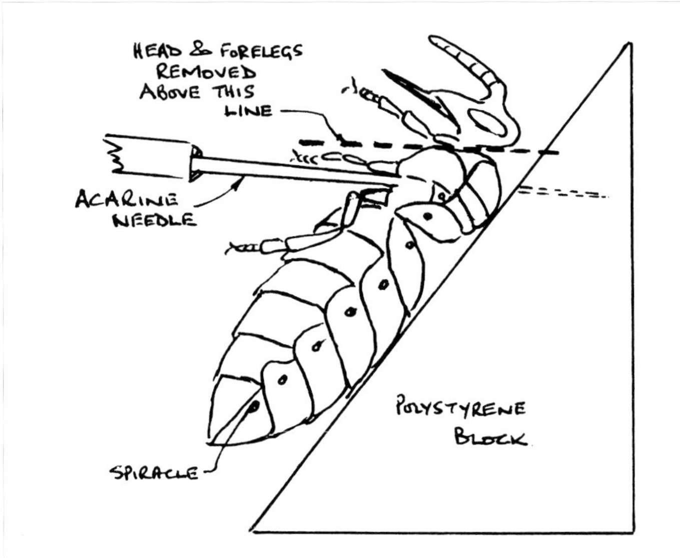

Diagram 15. The bee mounted for dissection and examination.

Whether you work left or right handed makes no difference; work whichever way you find most comfortable. With the round forceps hold the bee with its head to the top of the block and its wings to the surface of the block and pin it to the block with the acarine needle pushed through the thorax between the 2nd and 3rd pair of legs. The acarine needle should be virtually horizontal when the bee is in the correct position on a 45° block. The next operation is to grip the head and the 1st pair of legs in the round forceps and remove the head and legs with a twisting motion to the right or left and slightly upwards. All operations up to this point can be done without the microscope.

The bee mounted on the block can now be put under the microscope and the microscope brought into focus on the exposed portion where the head and legs have been removed. Note that this should be in a horizontal plane so that it is all in focus (hence the need for a sloping block of about 45°). In order to fully examine the trachea it will be necessary to remove the chitinous collar surrounding the exposed cavity. See diagram 16 below.

With the fine forceps grip the collar at 'A' and peel it off in a circular motion to 'B'. This is the part of the dissection which is the most difficult and practice will make perfect. Also, this is where a fresh sample is better than an old one. The target time needs to be a complete dissection and examination in 1 minute thus being able to complete a sample of 30 bees in 30 minutes. A competent microscopist must achieve this dexterity. The dissecting needle can be used to move any tissue obscuring the trachea from view.

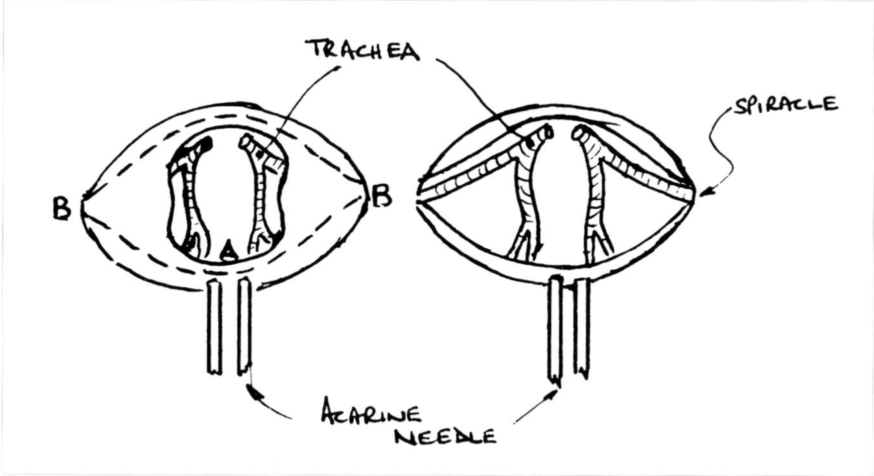

Diagram 16. Removal of the collar to expose the trachea.

Examination and diagnosis.

The whole of the 1st pair of thoracic trachea will now be visible up to the spiracle where they enter the exoskeleton. The trachea of a healthy bee should have a uniform creamy white colour. Where infestation occurs there will be a discolouration on either one side or both sides ranging from a slight bronzing of part of the trachea in the early stages of infestation to the whole trachea becoming black along its whole length. In between these two extremes dark specks or dark streaks will be visible.

The difficult part of the diagnosis is in the early stages when discolouration is just starting. However, with a good dissecting microscope and a magnification of x20 it is possible to detect the presence of individual mites before discolouration takes place. They appear as minute oval translucent bodies visible through the walls of the trachea. An easier and surer way is to dissect the trachea for examination on a slide. Candidates for this examination should be capable of removing one of the trachea and placing it on a slide to examine under higher power with the compound microscope. Any mites, eggs or nymphs will then be immediately apparent. In the first case examination is made with reflected light but in the case of the compound microscope (M=x40) examination is made with light transmitted through the specimen giving a much clearer image to observe.

Candidates are urged to have plenty of practice in dissection prior to the examination, there should be no doubt about a Candidate's dissecting dexterity.

When one bee has been found with an infested trachea in a sample of 30 bees it is hardly worth looking at any more as from the graphs there is a 99% probability that the colony has 20% of its population infested. However, assuming that the whole sample is examined and say 2 more bees were found infested then clearly this is an infestation of 3 bees in a sample of 30. This is, of course, 1 in 10 and from the graph it will be seen that there is a 99% probability that 47.5% of the colony is infested.

All the remnants of the dissection are put into the mortar for later use in the examination for nosema and amoeba.

Finally, we believe that association microscopists should make a recommendation about the treatment of acarine when it has been diagnosed in a sample. At the time of writing there is no recommended treatment for acarine despite the fact that we have treated in this country for about 60 years from the time it was discovered by Dr. Rennie. This is a matter that should be addressed by the NBU and the country advised accordingly. Our own investigations lead us to believe that Perizin could be the answer to the problem and it is readily available in the EU. However, since treating our colonies with synthetic pyrethroids for Varroosis we have seen very little acarine; we put this down to using these acaracides regularly every autumn.

4.2 The preparation and examination of a sample of bees for Nosema and Amoeba.

The apparatus required is as follows:

a) A compound microscope with a total magnification of between x300 and x600.
b) A pestle and mortar for preparing the sample. Other make-shift items made of glass or glazed crockery can be used such as an egg cup and the back of a teaspoon.
c) A small bottle with an eye dropper for adding water to the sample in the mortar. Distilled water is recommended in many books but tap water is just as good for this job. If you have doubts about tap water, examine a sample under the microscope and see what you can find - not a lot we suspect!
d) A supply of clean slides and cover slips. Slides can be effectively cleaned in a dishwasher if they are put into a suitable holder and washed in the vertical plane. Cover slips are so thin and fragile cleaning them for use a second time is somewhat difficult.

Preparation of the slide.

The method is easy. Using all the debris from the acarine dissections pound and mix the remaining bees in the pestle and mortar, adding a few drops of water at a time until a brownish coloured liquid appears at the bottom when the mortar is held at an angle of about 45° with the solid contents pushed to the higher side. Dab a clean slide with the tiniest bit of this solution about $^3/_4$ inch from one end. With the forceps place a cover slip carefully over the smear lowering it down slowly from one side to avoid trapping air bubbles. With the handle of the forceps press firmly down on the cover slip. Surface tension will keep the cover slip in place. There should be no liquid oozing out from the sides of the cover slip and the amount required will be quickly learnt with practice.

Examination of the slide.

The slide should be brought into focus at about x400 magnification. There will be plenty to see as follows:

a) Nosema spores 5µm to 8µm long shaped like grains of rice. They are quite characteristic and are unlikely to be mistaken for anything else. In the field of view the density of the spores can range from zero to 1000s. It should be noted that each spore has a double nucleus which appears as two tiny black dots in the rice grain. The microscope optics must be of a high quality and the instrument set up accurately to see the nuclei at x400 magnification.

b) Amoeba cysts are perfectly round about 10µm to 12µm in diameter. They can be confused with fat bodies and air bubbles to the inexperienced. The test is to rack the focussing and if the centre has a granular structure then it is a cyst. Fat bodies have a uniform texture throughout and air bubbles will have reflections from the surface of the bubble.

c) Pollen grains and pollen husks that have been digested by the bee are generally unmistakable once you start to learn about pollen recognition. There are many publications showing the shape and sizes of 100s of pollen grains found in the UK.

d) Plumose hairs are unmistakable and appear to be very large at x400.

e) Tracheae vary considerably in size down to about 0.1µm but are easy to recognise with unmistakable taenidia, the circular chitinous thickenings rather like the old gas mask tubes during the war. Try measuring the diameter of the 1st thoracic trachea on your microscope.

f) Malpighian tubules. These will be difficult to see as they are virtually without structure. However it is a good idea to dissect one complete for examination purposes and determine whether it contains Amoeba cysts.

g) Others. There will be pieces of body tissue, spent digestive cells, etc., etc.

Sometimes a disturbance in the fluid of the smear will be seen. It is caused by a protozoan flagellate (*Crithidia mellifica*), which is very small and transparent, swimming in the fluid between the slide and the cover slip.

The classification for the degree of infection can be determined as per the rule of thumb detailed in Section 4.7 and Appendix 2 under the classification of nosema infection into light, medium and heavy. If no nosema spores are found in the first field of view it is always instructive to search other fields of view to see if the odd spore can be found; usually one can be found as the disease is endemic in the UK. The degree of infestation is based on our experience using a microscope with a magnification of x400 with an NA = 0.65. If another microscope is being used with a different objective the spore count can be modified.

We believe that there should be a nationally agreed standard for the classification of nosema and the NBU could give a useful lead in this direction but seem reluctant to tackle this and other basic problems that affect all beekeepers. The use of a **micrometer net graticule** seems the simplest way by setting an arbitrary spore count in adjacent squares. A common micrometer net graticule is available for about £12, the same as an ordinary micrometer graticule, but engraved with a 10mm square divided into 400 smaller squares, ie. 20 x 20 with each smaller square measuring 500µm x 500µm.

Discuss with the Examiner:

4.3 The identification of worker bees with signs of Deformed wing virus or Chronic paralysis virus from specimens or images provided by the Examiner.

4.3 This is a new item in the syllabus and both Deformed Wing and Chronic Bee Paralysis Virus as the names suggest are caused by viruses which can only be seen with an electron microscope. It seems curious that the Examination Board have included this in the revised syllabus as it has nothing to do with microscopy. In our opinion this should be addressed in the Module 3 which addresses diseases of the honeybee.

In order to prepare Candidates for the examination the most we can do is to provide the best photographs that we could find, which accordingly are reproduced on page 41 below.

4.4 The identification of two pests from specimens, slides or images provided by the Examiner and discuss the anatomical features that enabled this identification. These pests will be selected by the examiner from the following list:

> **Varroa destructor, Tropilaelaps spp., Braula coeca and the larval, pupal and adult stages of small hive beetle, greater and lesser wax moths.**

This is a new item in the syllabus including further pests associated with the honeybee some which have not yet been found in the UK.

Varroa destructor: can be confused with ***Braula coeca***, see BSN Modules 1 to 3 Second Edition page 204 for a fuller description of both together with their physical size. Note that only the adult forms are required for the examination.

Tropilaelaps spp.: This is a mite which has yet to be found in the UK and again one must question whether it should be included in the syllabus together with the small hive beetle (*Aethina tumida)* both of which we have never seen. Again only the adult form needs to be recognised.

Similarly, as above, we have provided the most informative photographs that we could find of pests included in this section – see page 41-45.

4.5 The size of the sample required for the examination of adult bee diseases and its statistical significance.

Reference is made to the graph in Appendix 1 showing the percentage colony infection against sample size when one infected bee is found in the sample. The formula for drawing these graphs is also shown and was derived many years ago while the Authors were studying for the Senior Examinations. At that time we could find only sparse information on the subject in "The Illustrated Encyclopaedia of Beekeeping" by Morse & Hooper and in "Beekeeping Techniques" by Deans which are now out of print.

Using our formula and graphs give a higher percentage infection compared with the other two sources and since we are unaware of how their figures were derived we consider that they may be in error. Therefore, we prefer to use our own formula which we believe to be correct. The derivation is also shown in Appendix 1.

Without drawing the curves, or at least having them available, it would be very difficult to explain the statistical significance of the size of the sample required for examination of the adult bee diseases. The generally accepted sample size is regarded as 30 bees from the colony under consideration but this is not very critical. At least such a number will fit nicely into a matchbox! It will be noted that the graph flattens out quite rapidly with sample sizes greater than 30, so it can be argued that it is not worth examining a greater number to get a greater degree of confidence in the result. There is a much greater change in f_a from a sample size of 10 to 30 when compared with a change in sample size of 30 to 50.

Clearly the best sample, giving 100% degree of confidence, would be the total hive population which is impracticable and about 30 is a practical compromise after considering the mathematics involved. To understand what is happening consider various sample sizes of 10, 20 and 30 all of which have one diseased bee. From the graph there is a 99% probability that the colony has the following percentage infections, 47.5%, 28.7% and 20% respectively. Alternatively, consider some typical figures on the 80% possibility curve: At sample sizes of 15, 30 and 40 the percentage colony infection is 14%, 7% and 5% respectively.

4.6 How and where the adult bee sample should be taken from the hive and the reasons involved.

It is important to understand that we are examining bees that may have adult bee diseases and it is the old bees that are more likely to be infected rather than the young ones. A bee that has just emerged from its cell will NOT be infected with any of the adult bee diseases except maybe varroosis which is specifically excluded from our present considerations. BBKA Advisory Leaflet No102 "Taking a sample of adult bees", the only one that we know of, unfortunately does not mention this very important point.

'Old bees' is rather subjective and would be better defined as those bees that have completed their 3 week stint as house bees and have become foragers. These are the bees that are required for testing. Where is the best place to find them?

1. To be 100% sure then it must be those bees either leaving or returning to the hive and the sample can be taken from these but it is a bit tedious. Those leaving can be caught with a large polythene bag opened in front of the entrance and let 30 bees fly into it; count them on the way in. Alternatively, put a blanket over the entrance and wait until there are a sufficient number to scoop up in a matchbox. Both methods will ensure that only older bees are contained in the sample.
2. Young bees do not like the light and are likely to predominate in the centre of the brood nest. Old bees occupy the outer perimeter of the brood nest together with the returning foragers. Thus the first full

frame of bees working from the side walls of the hive is the next best place to take the sample. This is used by most beekeepers not having the time to fiddle around collecting them outside. Taking them from the first available frame is quick and can be combined with a normal colony inspection. Note that care should be taken not to include the queen.

There is one place where the sample should not be taken and that is over the feed hole with an inverted honey jar. This method has been, and still is, recommended in various books and leaflets. The feed hole in a crown board is more often than not above the middle of the brood nest just where the young bees will be found. There is the likelihood of obtaining the wrong bees in the sample using this method.

4.7 The assessment of the level of infection or infestation and likely outcomes if treatment is withheld.

It is understood that here we are addressing the two adult bee diseases Acarine and Nosema.

Acarine. See 4.5 above. The only way to assess the level of infection is from the graphs provided in Appendix 1. There is no authorised treatment for Acarine at the time of writing. Having said this, our experience leads us to believe that the use of acaracides for the treatment of Varroa has all but obliterated this disease together with the annihilation of the mite *Braula coeca*. We would suggest that if it is found in the autumn then treat with Bayvarol. If found in the spring wc would not trcat while supers are being added to the stock in question. If treatment is withheld the most likely outcome is that the stock will recover as it builds up on the spring flow and in the autumn is likely to survive the winter with the treatment.

Nosema. To the best of our knowledge, and despite our pleas going unheard by the National Bee Unit (NBU), the only assessment now is our classification into light, medium and heavy which we have evolved over a great number of years.

It is possible to examine a sample of 30 bees on an individual basis for nosema just as it is for acarine and then to apply the same reasoning from the graphs to determine percentage colony infection. The method would be tedious having to make up about 30 slides and then examine them individually. The more usual method is to prepare a homogeneous mix from the whole sample and make one slide with one smear.

When nosema is present those who are familiar with sampling for this disease, over a large number of samples, will readily appreciate the large variation in spore density from one slide (sample) to another. As it is now recognised that nosema is an endemic disease, it is highly probable that if one searched diligently enough it would be possible to find the spores in any colony. If this supposition is accepted, then one must ask the following question, "how many spores must be present before any treatment or corrective action is taken?" Unfortunately there is no mathematical or scientific way of resolving this problem as far as we know. The National Bee Unit (NBU) used to classify the infestation into light, medium and heavy but no longer do so; they use the expensive method of examining every bee. Any classification into light medium and heavy is purely empirical and based on practical experience. The Authors have for many years used the following classification for nosema when observing a random field of view on the microscope with an objective of 40 (with an NA of 0.65):

Continued on page 47.

Illustrations for section 4.3

The adjacent picture (attributed to the BBKA) shows bees with Chronic Bee Paralysis Virus (CBPV) with the typical K wings. It should be noted that there are two syndromes known, surprisingly, as syndrome one and syndrome two. Full details of the disease can be found in our BSN Modules 1 to 3, Second Edition at section 3.17.1 page 242.

The adjacent picture (attributed to the University of Florida) shows a newly emerged worker with Deformed Wing Virus (DWV). The signs are obvious from the photograph particularly the stubby useless wings with shortened rounded abdomen together with mis-colouring and paralysis. The life span is generally less than 48 hours. The condition is associated with *Varroa destructor*, the vector. For reasons unknown DWV is omitted in Module 3 but included here.

Illustrations for section 4.4

The adjacent picture shows a tropilaelaps mite to the right of a varroa mite. Both are easily distinguishable with a hand lens. There are two species namely *Tropilaelaps clareae* and *Tropilaelaps koenigerum* associated with Asian honeybees namely *Apis laboriosa, cerana, florea* and *dorsata.* The life cycle is similar to Varroa but slightly shorter thereby having a greater reproductive rate. The developed mite is reddish brown 1mm long and 0.6mm wide.
DEFRA have produced a very informative leaflet as a pdf which can be downloaded from the internet.

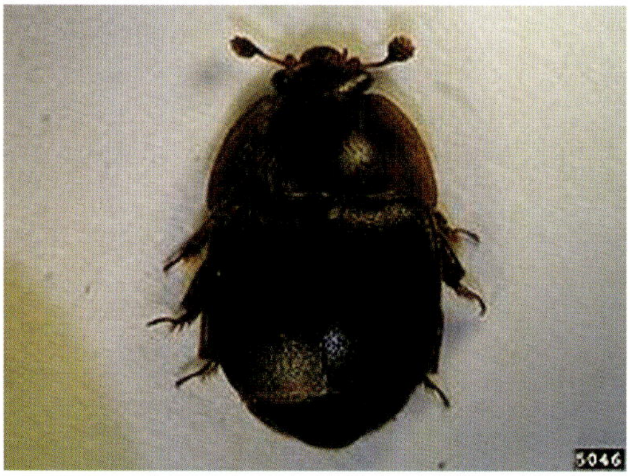

Small hive beetle (*Aethina tumida*), adult form, is shown in the adjacent picture.
Starting from the egg stage, each egg is 1.4mm x 0.26mm and pearly white in appearance and take 2/3 days to hatch into a larvae. The larvae grow to a length of 10 to 11mm long with 3 pairs of legs near the head with small spines along the dorsum and are coloured beige. They feed on pollen and honey for a period of 10/14 days. Then the larvae crawl outside the hive to pupate in the adjacent soil and have a pupation period of 3/4 weeks. 4/5 generations can be produced in one year. The adult (5 to 7mm long and 3 to 4.5mm wide) is of reddish brown colour which darkens to a brown black colour as the beetle ages. The females can lay up to 2000 eggs during her lifetime so only a few beetles can produce a severe infestation. The adults can fly up to 16km in search of honeybee colonies. See FERA's pdf named Small_hive_Beetle_Feb_2011.pdf which can be down loaded from the net. The two pictures below show the larval and pupal stages.

Greater and Lesser Wax Moths.

It is to be noted that the syllabus requires the Candidate to be able to recognise the larval, pupal and adult forms of these wax moth pests. The life cycle of the greater wax moth (*Galleria mellonella)* may be found in our BSN edition 2 Module 3 page 260. However the life cycle of the lesser wax moth *(Achroia grisella)* could not be found in all the books we searched at that time. We have recently widened our search with the 'Good God Google' and believe that we have stumbled on the only data available on the subject attributable to, we believe, the University of Florida. and is shown in tabular form below:

General characteristics of greater and lesser wax moth life stages. **from:** http://www.coloss.org/beebook/II/wax-moth/2 attributed to the University of Florida.		
	Lesser Wax Moth Larvae	**Greater Wax Moth Larvae**
size	1 - 20 mm long; fully grown = 18.8 ± 0.4 mm (length)	first instar = 1-3 mm (length) fully grown = 12-20 mm (length), 5-7 mm diameter
description	narrow white bodies with brown heads and pronotal shields	creamy-white with grey to dark grey markings, a small slightly pointed, reddish head
length in life stage	6 - 7 weeks at 29° to 32° C; 30.10 ± 2.5 days	6-7 weeks at 29° to 32° C2
diagnostic characters	Stemmata absent; spiracle with black peritreme thicker on caudal margin	Head with 4 stemmata on each side; spiracle with yellowish peritreme of uniform thickness
	Lesser Wax Moth Pupae	**Greater Wax Moth Pupae**
size	11.3 ± 0.4mm in length & 2.80 ± 1.89 mm in width	12-20 mm in length & 5-7 mm in width
description	yellow-tan pupa in a white cocoon often covered in frass and other debris	dark reddish brown pupa in an off-white, parchment-thick cocoon
length in life stage	37.3 ± 1.2 days	6 - 55 days depending on environmental conditions
	Lesser Wax Moth Adults	**Greater Wax Moth Adults**
size	male = 10 mm long female = 13 mm long	15 mm (length) with a 31 mm average wingspan
description	small, silver-bodied with a conspicuously yellow head, oval shaped fore-wings and heavily fringed hind-wings	heavy-bodied, reddish brown with mottled fore-wings and pale cream-coloured lightly fringed hind wings
lifespan	female = 6.90 ± 1.135 days male = 12.90 ± 1.30 days	female = ~ 12 days male = ~ 21 days
diagnostic characters	Fore-wing breadth less than 5 mm; termen of fore-wing convex (hind-wing	Fore-wing breadth 5 to 7 mm; termen of fore-wing concave; Cu of hind-wing

It is to be noted that there seems to be some discrepancies in the "diagnostic characters" of the adults which we couldn't determine. This, however, does not detract from the usefulness of the tabulated information.

The greater wax moth has, in our experience become a greater pest than formerly when para-dichlorobenzene PDB was readily available particularly for storing brood comb during the winter. It was banned by EU instigated regulations on the grounds that it was carcinogenic (encourages the growth of cancer) although the US Department of Health and Human Services state that it may be reasonably anticipated to be carcinogenic, there is no direct evidence of such. We have never heard of anyone being this afflicted by using PDB in this country!

We have for some 50 years always stored our supers each winter with PDB - a level dessert spoonful sprinkled on newspaper below each super. By the spring the PDB crystals have evaporated and they went straight back on the hives. There was no detriment to the bees or the new crop of honey stored in them.

We now use look alike moth balls sold for protecting clothes. So far we have not had trouble with stored supers but we have had trouble with brood comb. We have tried Certan with little success and find it a very expensive and inefficient solution for preventing damage by wax moths.

The picture above, is attributed to the University of Florida and shows the adult form of both types of wax moth (M=male and F=female).

The adjacent picture, again taken from Google pictures on the net, shows the larval and pupal stages of the greater wax moth together with the silk cocoon before the adult moth appears.

We are a little bit suspicious of the colouring in this picture as the larva should appear as creamy-white with grey to dark grey markings, a small slightly pointed, reddish head.

The pupa colouring appears to be reasonably correct, dark reddish brown pupa in an off-white, parchment-thick cocoon.

We have not been able to find a similar picture for the lesser wax moth, the details of which seem allusive as ever.

Illustrations for section 5.3

Chalk Brood - In the comb (above left) and Out of the comb (above right)

Pollen pellets are shown in the adjacent picture. note the distinctive difference in colour between Chalk Brood mummies and pollen pellets. chalk Brood mummies are always white/grey/black in colour. There are only a minority of pollens grey in colour and none, to the best of our knowledge, that are white or black with the exception of poppy pollen which is black.

Colours of the hazard symbols used in the appendices 4 and 5.

To keep the cost of production of this book to a minimum all colour plates have been printed in a single batch of coloured pages (41 to 46). The symbols in the two appendices referred to above have been printed using the grey scale from black through to white. Here we show these symbols in the colours that they would appear on labels.

Figure 2 The seven familiar DSD/DPD symbols **The nine GHS pictograms**

Figure 2 above shows the old system with its typical orange coloured hazard symbols where DSD = **D**angerous **S**ubstance **D**irective and DPD = **D**angerous **P**reparations (ie. mixtures of chemicals) **D**irective. These were mutually agreed throughout the world.

By 2015 it will be mandatory in the EU to use the new EU derived symbols in red, black and white called GHS pictograms where GHS = **G**lobally **H**armonised **S**ystem of classification. It is puzzling why they were changed in the EU but the clue may be in the name using the word 'harmonise' - everything to do with the EU must be harmonised.

* ** *** ** *

Section 4.7 *Continued from page 40.*

Light	0 to 5 spores (using x400 magnification)	
Medium	6 to 15 spores	(ditto)
Heavy	>15 spores	(ditto)

The old treatment for these nosema infections was as follows:

In the spring:
(light and medium) - FUMIDIL B.
(heavy) - FUMIDIL B plus Bailey frame change and re-queen as soon as possible. (We find it more expedient to kill off the colony as it seems to take the whole season 'monkeying around' with it to bring it back to normality.
In the autumn:
(light, medium and heavy) - FUMIDIL B with the winter feed.

Using this basis for managing our apiaries over many years, we have only lost one colony in the winter that was attributable to nosema. We find about 10% of our colonies need treatment for nosema each year and regard the disease as a greater management problem than acarine in our part of the world. Our own association microscopists are using the same criteria for samples sent in by members and the form that they use is reproduced in Appendix 2 which when completed is sent back to the beekeeper originating the sample. There is no known treatment for Amoeba.

Finally, the likely outcome if treatment is withheld:

In the spring - the colony will dwindle and be an unproductive unit for the whole season.
In the autumn - the colony is likely to succumb during the winter period.

4.8 What advice should be given to beekeepers on the actions to be taken and sources of information in the event of an adult bee disease or colony infestation being identified.

This is a new item in the syllabus and is a classic example of imprecise wording by the Examination Board, vis. "or colony infestation being identified". Are the diseases Chalk Brood, Sac Brood, Varroa, the Foul Broods, etc. to be addressed? Heaven knows! We shall assume that only Acarine and Nosema are to be addressed and these are fully covered in item 4.7 above.

If other diseases are meant to be included then the most concise source of information on them all is contained in our BSN Modules 1,2&3 Second Edition, Module 3 pages 180 to 263 which we consider to be essential reading before attempting this Microscopy Examination.

* ** *** ** *

5. DIAGNOSIS OF AFB AND EFB

Discuss with the Examiner:

5.1 The key features of American and European Foul brood including the difference between healthy and diseased larvae.

This is a new item in the syllabus and is fully covered in BSN Modules 1, 2 & 3 Second Edition, Module 3 pages 180 to 196.

5.2 What advice should be given to beekeepers on actions to be taken / sources of information in the event of AFB / EFB being identified.

As above this is a new item in the syllabus and is fully covered in BSN Modules 1, 2 & 3 Second Edition, Module 3 pages 180 to 196.

5.3 The identification of chalk brood from specimens or images provided by the Examiner.

This is a new item in the syllabus and it would be advisable to read up on Chalk Brood in BSN Modules 1, 2 & 3 Second Edition, Module 3 page 230. Do not be caught out by confusing a Chalk Brood mummy with a pollen pellet. Both are illustrated on page 45.

5.4 More on AFB and EFB slides which has been deleted from the syllabus.

The previous syllabus included the diagnosis of AFB and EFB and we believe that it is a matter of some importance and, therefore, we include the script from the old second edition below.

5.4.1 How an authorised person would obtain a suitable sample for AFB from infected comb.

If AFB is present in a colony of honeybees the first indications are likely to be the usual signs of the disease, the characteristic appearance of the cappings and a pepper pot brood pattern. At such a stage the unused cells will exhibit the characteristic tell tale signs; the hard black scale in the lower angle of the cells which the bees cannot remove. To diagnose AFB a small piece of this scale is required about the size of a couple of pin heads which will contain thousands of the spores of *Paenibacillus larvae larvae*. The total scale in each empty cell will contain about 2,500 million spores.

A suitable frame should be shaken free of bees and removed from the colony. The method of removal is important. There must be no bees left on the frame and it must be put into a bee-proof container or box to remove it from the apiary, thereby denying access to the frame by any other bees.

Once the frame is taken indoors it should be opened in a bee-proof room and a piece of scale removed from the cell by the use of forceps and placed at one end of a microscope slide ready for preparation. (Note that the forceps will require to be sterilised before and after use.)

It must be emphasised that the frame removed for obtaining the sample will have to be destroyed in accordance with the statutory requirements if AFB is subsequently confirmed.

5.4.2 How to obtain a sample for EFB and the appearance of an infected larva.

Obtaining the sample for EFB also requires the removal of a frame and it must be treated in the same way as the frame for AFB and regarded as highly infectious. It must be destroyed or replaced in the colony depending on whether the colony is to be destroyed or treated if the disease is present. The selection of the frame is a bit more difficult for EFB because larvae that are about to die are the ones that are required. This involves being able to recognise a larva of the right age and one that is displaying the characteristic signs of EFB; that is, discolouration and distortion. In practice it is best to remove a frame with larvae spanning a range of ages from day old to the oldest that are unsealed, noting that infected larvae are often removed by the bees.

When the frame is examined indoors, again in a bee-proof room, larvae should be removed carefully until one is found that has a white appearance of its ventriculus. This white mid-gut will be quite evident when compared with a healthy larva, which has a ventriculus with a brownish tinge. In a badly infected larva, the white appearance is clumps of millions of EFB spores of *Melissococcus plutonius* inside the ventriculus. The selected larva should be placed at one end of a microscope slide ready for preparation.

5.4.3 How each sample would be processed and the smear prepared for examination under the microscope.

The equipment required for preparing slides of the two diseases is as follows:

- small platinum wire loop in holder (loop c. $^1/_8$in diameter)
- 2 pairs of fine forceps
- Bunsen burner or small portable gas torch
- black stain (nigrosin)
- Canada balsam
- Cedar oil
- 1in x 3in microscope slides
- thin cover slips

5.4.3.1 AFB slide. The starting point is the piece of AFB scale on the end of a 1in x 3in microscope slide as discussed in section 5.1 above. The procedure is as follows:

1. Sterilise the platinum loop in the burner flame until it is a cherry red colour.
2. Mix the piece(s) of scale sample with a drop of water using the platinum wire loop.
3. Transfer part of sample to the end of another slide, the test slide. The loop now has millions of spores on it and must be sterilised again.
4. Heat the loop in the burner until it glows a cherry red colour.
5. Place a drop of stain on the original slide adjacent to the sample with the wire loop.
6. Mix a little stain with the smear on the test slide. The colour should be a dark grey; colour is always difficult to describe in words and practice will make perfect!
7. Sterilise the mixing loop.
8. Dry the smear on the test slide in the flame of burner. We prefer to put the smear at one end of the test slide, rather than the traditional way in the middle, in order to leave most of the slide for labelling.
9. There are two ways to proceed to examine the test slide; one with a cover slip (permanent) and one without a cover slip (temporary). If the slide is viewed without a cover slip and is a positive result, a second one can be made with a cover slip from the same prepared material knowing that the staining is satisfactory. For the temporary slide put a small drop of cedar oil on the dried smear and view at c. x1000 diameters.
10. Take great care bringing the objective into contact with the cedar oil by viewing from the side of the microscope and only looking down the eyepiece when this has been done. Only the fine focussing screw on the microscope should be used as the distances are very small to adjust the focus. If AFB is present the *Paenibacillus larvae larvae* spores will be randomly distributed over the field of view and look very similar to nosema except, of course, for the size which is c. 1.0μm x 0.5μm.
11. If a permanent slide is to be made proceed as 1 to 8 above and then put a small drop of Canada balsam on the smear followed by a thin cover slip. Everything should be warm before lowering a warm cover slip onto the Canada balsam with the forceps to exclude air bubbles.
12. Leave for 24 hours at room temperature for the Canada balsam to dry and then ring with nail varnish.
13. View with cedar oil on the cover slip as in 9 above; an example of negative staining.

Do not forget to clean the microscope objective (See section 2.7.3) and pay particular attention to cleanliness when dealing with foul brood material. The platinum loop should always be put down after use in a sterile condition; failure to observe this routine will contaminate the work area with foul brood spores. Cover slips or glasses come in a variety of thicknesses which are approximately as follows:

Thin - 0.14 mm   Standard - 0.17 mm   Thick - 0.20 mm

The thin variety is essential when using high magnification with oil immersion because of the short focal length of the objective and its proximity to the cover slip when in focus.

5.4.3.2 EFB slide. The starting point here is the larva at one end of the slide as discussed in section 5.4.2 above. The larva is pulled apart with two pairs of forceps and the white mid-gut used as the sample. Proceed making the slide as for AFB up to stage 8 as above. Rather less water will be required as the mid-gut with the peritrophic membrane is quite moist; again, practice makes perfect.

EFB spores, *Melissococcus plutonius*, are about the same size (1.0µm x 0.5µm) and are diamond shaped with the four corners rounded, not very different from the rice grain shape of AFB. However, the distribution in the field of view is very different and not random as before. The individual spores will appear in long strings and / or in floral patterns where the strings cross one another.

5.4.4 Other points.

It is noted that the old syllabus stated that Candidates will not be required to diagnose an actual sample because of the statutory requirements involved. However, there is nothing to stop any beekeeper making his own diagnosis if he suspects AFB providing he observes the statutory requirements to prevent the spread of the disease.

It should also be noted that it is possible to see both AFB and EFB with a microscope that is well set up using x400 magnification without the oil immersion.

* ** *** ** *

6. POLLEN IDENTIFICATION

Discuss with the Examiner:

A few preliminary observations about pollen grains and the terms used.

Palynology is the analysis of pollen grains found in ancient sediments and soils. **Pollen identification** enables beekeepers to discover where bees are foraging for nectar and pollen.

Pollen grains are produced in the anthers of the stamens of flowers. When the anther is 'ripe', the outer layers of the anther dry out, tension is set up in the lateral grooves and eventually 'dehiscence' occurs. This is when the pollen sacs of each anther burst open, the outer layer of the pollen sacs curl away exposing the pollen grains. Pollen grains carry the male reproductive cells of plants. See diagram 17 below.

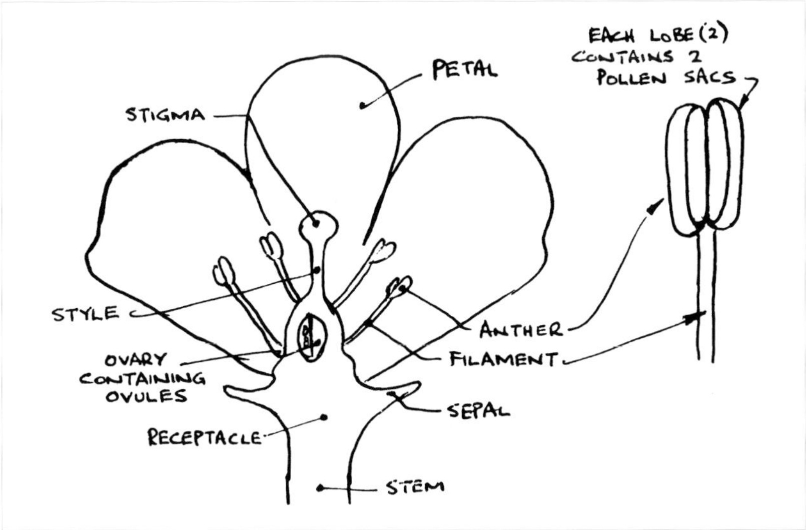

Diagram 17. Section through a generalised angiosperm flower.

There are two main ways in which pollen is transferred to a receptive stigma:

1. Anemophilous flowers are wind pollinated, these pollen grains are smooth walled, small, dry and light and are easily carried in air currents. Anemophilous flowers are characteristic of trees and grasses eg. a single catkin of the silver birch tree, *Betula pendula*, has over 10 million pollen grains.
2. Entomophilous flowers are pollinated by insects, smaller amounts of pollen grains are produced in fewer anthers, one anther of the Red Clover, *Trifolium pratense,* contains over 200 pollen grains. The pollen grains are larger, often have sculptured walls and adhere easily to the exoskeleton of visiting

insects. Some entomophilous flowers have highly specialised pollination mechanisms which prevent self-pollination and promote cross pollination.

Other agents of pollen transfer are water, birds and bats.

Pollen identification of grains taken from the corbiculae of bees entering the hive will confirm the species of plant from which the pollen load originated. Bees usually forage within a 3 mile radius of the hive. Knowledge of the local flora and the colour of the pollen loads will suggest to the observer where the bee may have been foraging but only microscopic examination of the grains in the pollen pellet will give positive identification of the flowers visited by the bee.

Melissopalynology (pollen analysis of honey) will reveal the origins of the nectar from which the bees have processed the honey. The amount of pollen found in a sample of honey will depend on the methods used to extract and filter the honey from the combs.

Absence or lack of pollen may be due to:

a) The filtering of honey through diatomaceous earth filters under pressure.
b) Honeys derived from nectar collected from either extra floral nectaries, dioecious plants which are female and produce no pollen or plants which do not disperse their pollen eg. *Orchidaceae.*
c) Honeydew honey which is wholly derived from the exudate of aphids feeding on the sap of plants. Honeydew honey may contain small amounts of anemophilous pollen and no entomophilous pollen.
d) Inappropriate feeding of sugar syrup or adulteration of the honey with high fructose corn syrup.

Uses of professional analysis of pollen grains:

All honeys for sale must be correctly labelled. Misrepresentation of honeys can be detected by the presence of unexpected pollen grains. Imported honeys are usually cheaper then English honeys. Unscrupulous honey packers may attempt to sell these imported honeys as English honey. Identification of exotic pollen grains only found in imported honey eg. the large pollen grains of the eucalyptus tree, *Banksia spp.,* will confirm the foreign origin of the honey. Similarly the absence of the tetrad pollen grains of ling, *Calluna vulgaris,* in a jar of honey, labelled 'Heather Honey', will indicate that honey has been misrepresented by the honey packer in order to obtain a higher profit.

In 1992 identification of parsnip pollen by a botanist at Kew was used as evidence by Essex beekeepers successfully prosecuting a spray contractor for loss of colonies of bees. In this incident a 33 acre field of parsnips being grown for seed had been sprayed with Hostathion and Dimethoate during daylight hours whilst the bees were foraging on the flowers. Both containers of the pesticides had carried warning labels saying: "Dangerous to bees. Do not apply at the flowering stage".

6.1 The general construction of a pollen grain.

6.1.1 General.

The pollen grain is a living cell surrounded by two protective coats called **intine** (interior) and **exine** (exterior). The intine is a semi-permeable membrane which does not absorb staining. The intine may bulge through the apertures in the exine eg. in the family *Caryophyllaceae*. Within the intine are the cytoplasm and two nuclei. The outer tough exine is made up of two layers. An inner homogeneous layer and a complicated outer layer on to which various patterns of rods, spikes, tectum (cover) or other ornamentations may appear. See diagram 18.

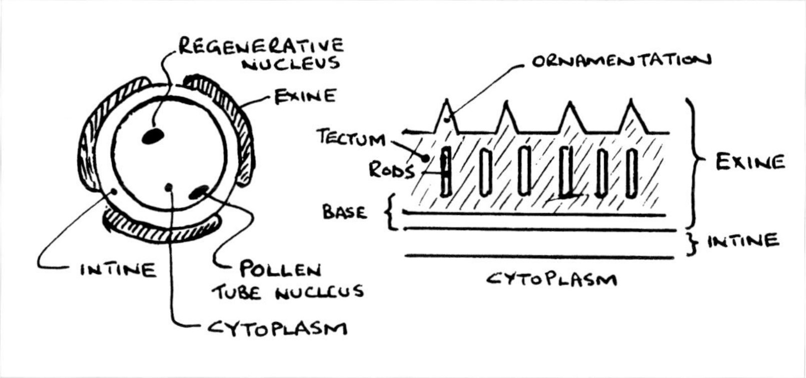

Diagram 18. Mature pollen grain and the grain wall.

Apertures on the exine allow the pollen grain to dry or to absorb water and expand. Apertures are formed where the exine has become thin or are totally absent. There are two main types of apertures called *furrows or colpi* if elongated or *pores* if round in appearance. When a compatible pollen grain is deposited on to a receptive stigma the pollen grain germinates. A pollen tube grows out from the pollen tube nucleus through one of the apertures. The tube penetrates the stigma and grows between the cells of the stigma and style towards the micropyle of the ovary. The generative nuclei will divide into two male gametes which pass down the pollen tube into the embryo sac. One male gamete will fuse with the female nucleus to form a zygote and the other nucleus fuses with the endosperm nucleus. This double *fertilisation* is a unique feature of the angiosperms.

Pollen grains which have been collected by honeybees lose their power of germination. It is thought that this is due to the fatty acid (10-hydroxy-2-decenoic) from the mandibular glands of the honeybee. When the pollen is packed into the corbiculae by the foraging honeybees, secretions from the honey sac are mixed with the pollen load. When the pollen is stored within the hive further honey is mixed with the pollen, it is packed into the wax cells and finally sealed with a layer of honey and capped over with wax.

Pollen grains are a rich source of food, especially of proteins:

protein	16 - 30%
starch	1 - 7%
sugars	0 - 15%
fat	3 - 10%
ash	1 - 9%

plus water, enzymes, minerals and vitamins.

Honeybees are stimulated to forage for pollen by pheromones given off by the queen and the presence of open brood. Honeybees are able to discriminate between pollens by colour and odour but are unable to distinguish the quality (protein content) of various pollens.

The exine of the pollen grain is very resistant to decay. Pollen grain features vary from species to species though pollen grains from the same genus will have similar characteristics. These differences may be listed under *size, shape, aperture number and type, surface pattern, composition of exine, colour and any other external structural irregularities.*

6.1.2 Size.

The size of the pollen loads collected by bees varies from plant to plant. For example, pollen loads from clover, *trifolium repens* are always small and compact whereas loads from the dandelion, *taraxacum spp.* or charlock, *sinapis arvensis* are large covering the tibial segments of the hind legs and part of the basitarsi. Pollen is also carried back to the hive on the exo-skeleton of the bee. The plumose hairs on the exo-skeleton of the honeybee trap pollen grains whilst the bee is working the flower for nectar or pollen. It is a common sight when bees are working the dandelion blossom to see the returning foraging bees covered with a 'yellow dust'. The size of pollen grains may vary from about 6µm eg. forget-me-not, *myosotis spp.,* to about 140µm eg. the hollyhock, *althaea rosea*. Most pollens fall within the range 20µm -30µm [one micron equals 10^{-6}m or 1/1000 millimetre].

The maximum diameter of the pollen grain including spines is measured in micro-metres (µm). Size is an important aid to identification. Using an eye piece of power x10 with a graticule and with an objective of power x40 (magnification = x400) each division of the graticule is equal to c. 2.5µm. Some pollens grains within the same genus are very difficult to identify. Size may vary with pollen obtained from the same plant eg. In the *Rosaceae* family the pollen is so similar in shape that species are grouped together by size eg. fruit blossom, cherry / pear, if the pollen is over 30µm or blackberry / raspberry, if under 30µm. Correct identification of pollen grains is in some cases very difficult. A great number of pollen grains measure between 20µm -30µm diameter. At times it may only be possible to identify the species or even the family.

Factors affecting the size of the pollen grain:

a) The size of the pollen grain enlarges due to swelling of the exine when the pollen grain absorbs the mountant. This process makes identification of the pollen grain easier as the external characteristics of the exine become more pronounced.
b) The size and shape of the pollen grain will vary depending on the optical aspect being viewed eg. polar or equatorial view. The polar areas are the two opposite areas with no apertures. The equatorial area is midway between the two poles. The furrows will run longitudinally with the narrow ends towards the poles. See diagram 19 below.
c) The size of some pollen grains will vary within the same species. This may be due to nutrition or variety eg. pollen taken from Winter heather, *Erica carnea,* in the wild may be smaller than a cultivated variety found in a garden. Purple Loosestrife, *Lythrum salicaria,* has trimorphic flowers which yield different size pollens.

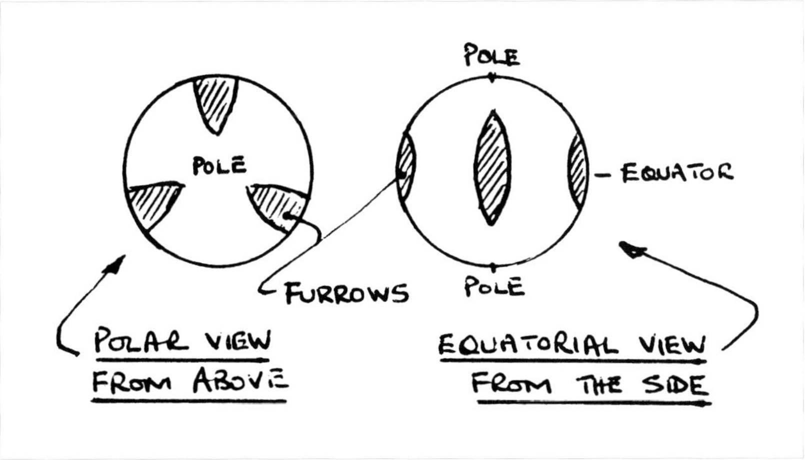

Diagram 19. Pollen grain with 3 furrows.

6.1.3 Shape and structure.

The shape or outline of pollen grains will vary from species to species. The shape, depending on the aspect from which it is viewed under the microscope, can usually be placed into one of 7 categories.

a) Round eg. Crocus, *Crocus aureus.*
b) Oval, flattened eg. Indian Balsam or Policemen's Helmet, *Impatiens glandulifera.*
c) Oval, elongated eg. Buckwheat, *Fagopyrum esculentum.*
d) Triangular eg. Lime, *Tilia vulgaris.*
e) Long eg. Hogweed, *Heracleum sphondylium.*
f) Semicircular or boat shaped eg. Ransom or wild garlic, *allium ursinum*
i) Irregular or multi-sided eg. Ling, *Calluna vulgaris.*

6.1.4 Aperture Numbers and Type.

The number of apertures can be counted, sometime only pores can be seen, sometimes only furrows or there may be a combination of both types, colporate. Sometimes the pores or furrows will appear to be encrusted with warts eg. the pollen of the Horse Chestnut tree, *Aesculus hippocastanum.* Some of the grains require a higher magnification than 400 to reveal the openings in the exine eg. the pollen of Meadow Foam or Butter and Eggs, *Limnanthes douglasii.* Abnormalities occur especially in the *Rosaceae* family where 4 furrows appear amongst the 3 furrowed varieties.

6.1.5 Surface.

Various terms are used to describe the surface of the exine but the simplest classifications are smooth, granular, striate, net or net beaded and isolated dots due to spines or projections. Pollen grains are semi-transparent. By adjusting the fine focus of the microscope and moving the 'focal plane' up and down, 'racking' through the grain of pollen on the slide, the pattern of the exine can be observed.

6.1.6 Other structural differences.

These differences are many and various for example pollen from Pine trees, *Pinus sylvestris,* consists of a single grain with two air sacks. The pollen grains from Fuchsia, *Fuchsia magellanica,* have viscous threads attached. Compound grains like the tetrads are found in the *Ericaceae* family. Sometimes a giant grain will appear or an extra furrow or pore present itself. Some pollen grains remained shrivelled after mounting, this is especially noticeable amongst pollen collected from the cultivars for reasons unknown to us.

6.1.7. Colour.

The colours of pollen grains vary considerably. Most common are the shades of yellow or orange. Colour is an unreliable guide to the correct plant species, especially when the flowering period of species whose colour range is similar coincide eg. apple, pear and plum. Other variations of colour may be caused by:

a) Grains are darker in the early morning, after frost or rain.
b) Lighter coloured grains are found during fine weather and in the latter part of the day.
c) The colour of the pollen may vary from season to season.
d) Some pollen grains will change colour with age or with moisture content.
e) Within the same species the colour of the pollen may vary as in the Purple Loosestrife, *Lythrum salicaria,* the plants are trimorphic. Bottle green pollen comes from the long stamens on short styled flowers. Flowers with long styles and both medium and long stamens with white filaments produce yellow pollen. Lastly, if the flowers have a medium size style and long stamens with red filaments the anthers produce emerald green pollen.
f) Discoloration due to adulteration of the pollen by soot or dust particles.

g) The foraging bees collect, moisten and pack the pollen in their corbiculae. The hive bees pack and preserve the pollen within the hive. These pollen grains will vary in colour with the type and quantity of honey which is used.
h) Honeybees are polytropic i.e. they will forage on many species of flower. However, honeybees are said to remain constant to one species of flower, this trait makes them reliable pollinators. Nevertheless about 1% of the bees foraging for pollen will return to the hive with mixed pollen loads.
i) Some pollen grains are easily identified by colour eg. the black pollen of the Poppy, *Papaver rhoeas,* the bluish green pollen of Rosebay Willow Herb, *Epilobium angustifolium* or the bright orange colour of the Dandelion, *Taraxacum officinalis.*
j) Unprepared pollen grains for microscopic examination are in a dry shrunken condition. Preparation requires the grains to be first cleaned with ether or alcohol to remove any oil. The sample of pollen grains is allowed to dry. A small amount of warm glycerine jelly is added to the pollen. The exine, the outer covering of the pollen grain, absorbs the reagent and becomes thin and colourless.
Note the colour of the pollen grains differ from the anther which produces it.

6.2 The collection and preparation of pollen from (a) flowers, (b) pollen loads from the honeybee and (c) honey.

Equipment required:

The basic equipment required is a dissecting set which includes the following: glass slides, cover slips, large and small forceps, small sharp scissors, pipettes long and short, test tubes, scalpel, needles, glass rod, ether or alcohol, spirit lamp or small hot plate, mounting agent (Fuchsin glycerine jelly), colourless nail varnish for sealing specimens. A mechanical stage will be required on the microscope for counting the concentration of types of pollen.

(a) Flowers.

To be certain of obtaining a 'true' sample of pollen directly from a flower, the collector should pick a sample of the flower in bud from an identified species of plant, put the stem into water and keep the specimen in a warm room with plenty of light, say from a window, in order that the flower shall open naturally and remain free from contamination.

a) Cut a sample of the flower in bud, early morning, and bring into a warm closed room.
b) Wait until the anthers are mature and ready to dehisce.
c) Remove the flower from the stem and cut out the stamen onto a glass slide with small sharp scissors or razor blade.
d) Shake or remove with needle some of the pollen from the anther on to the middle of the right hand side of the slide. This will leave the left hand space for labelling of the specimen. Avoid any contamination from previous work.

e) Cover the pollen with a drop of ether / alcohol to remove any oil. This procedure can be omitted but the pollen grain if not degreased may not take up the staining. When the ether has evaporated, clean any residual debris away from the slide or place the pollen on a clean slide.
f) Place a small drop of warm glycerine jelly containing Fuchsin over the pollen, keep this warm, not over 40°C, for about five minutes to allow the pollen to swell and absorb the stain, some pollens will take longer to swell.
g) Using forceps carefully place a warmed cover slip over the pollen, lower the left hand side first and carefully cover the specimen. Try to avoid incorporating any air bubbles. Now you will know if you have used too much glycerine jelly as this should not come to the edge of the cover slip.
h) Check you have made a good specimen before sealing with colourless or coloured nail varnish which is easy to use. Melted paraffin wax applied with a glass rod to the edges of the cover slip whilst the slide is still warm is recommended by some authorities but in time the wax will harden and crack. The seal of wax is more easily damaged and is not used by the Authors. Round cover slips are more expensive but easier to use. Thickness of the cover slip is not critical for pollen slides which will not be viewed using oil immersion.
i) Measure and identify the pollen grains.
j) Now label your specimen with date, family, genus, species and size. File and store in a slide rack in a cool dark cupboard. With age the pollen may swell and fade but it will remain a useful reference for many years if well sealed.

(b) Pollen loads from the honeybee.

Collect pollen pellets for analysis by either gathering the discarded loads of pollen which are often to be found on the alighting board during a nectar flow, use a pollen trap for a short period, trap the returning foraging bees with pollen loads and remove the pollen pellet from the hind legs with a wet camel hair brush (no 1 or 2) or lastly during colony inspection pellets can be picked up with a pin. Store the pellets in a clean envelope.

a) Label the envelope with the date and the source of the pollen. Keep the pollen pellets in a cool clean dark place until you are ready to make your slides.
b) Using the glass rod mix the pellet with a little ether / alcohol on a clean white tile or plate to de-grease the specimen and allow the alcohol to evaporate.
c) Add a little distilled water to the pellet and mix with the glass rod.
d) Using the glass rod make a smear onto a glass slide and allow the smear to dry.
e) Make your slide as before, using only a little of the pollen.
f) Identify the pollen, seal and label the slide.

(c) Honey.

Unless honey has been treated under pressure through a very fine filter, all floral honeys will contain pollen grains. Two methods, centrifuging or sedimentation, can be used to obtain pollen grains from samples of honey in order to determine the geographical origin of the nectar collected by the bees.

In the United Kingdom there are a few mono-floral honeys obtained when hives of bees are moved directly onto the crop eg. heather, clover, rape or field beans. On the supermarket shelves imported honey can be found labelled as being derived from Orange, Lavender or Acacia blossoms. Under the Honey Regulations applying to the sale of honey in the UK this honey must be derived wholly or mainly from the blossom of the plant indicated. County Trading Officers can submit samples of honey to the public analyst for microscopic examination. Prosecution will follow if care has not been taken to observe the Honey Regulations and other legal requirements, laid down by the European Union, covering the sale of honey. The examination in a laboratory would require a high degree of accuracy if a prosecution is to be made for misrepresentation. Calculating the pollen composition of a honey in terms of its floral sources is outside the scope of this syllabus.

The following tests can be carried out by the beekeeper and the results would show what flowers have been visited by stocks of bees to produce the harvest of honey. In a sample of honey there will be more of the smaller pollens eg. Forget-me-not or Sweet Chestnut. These pollens are said to be over represented. Honeydew honey or honey contaminated with honeydew eg. Lime may also contain pollen from amenophilous plants eg. grasses, *Graminae*, characteristic small square crystals, algae, fungal spores and mineral dust.

Centrifuging Method.

Warm the honey to be sampled and give it a good stir with the glass rod.

a) Place 10g of honey in a 50ml beaker (or use a clean honey jar). Weigh the sample using electronic kitchen scales with a 'tare' facility.
b) Add water to make a 30ml solution.
c) Divide the solution into two test tubes, 15ml each and centrifuge for 5 minutes at 2,000 revolutions a minute.
d) Decant the water from both tubes leaving the sediment behind.
e) Half fill one tube with water, give it a good shake and add the solution to the other sediment.
f) Fill the empty tube with water to the same level as tube with the sediment to balance the centrifuge rotor arm.
g) Centrifuge again for 5 minutes.
h) Decant the liquid from the tube with the sediment.
i) Stir the sediment with a long pipette. Draw up the sediment into the lower end of the pipette and spread the sediment onto a microscope slide
j) Dry the sample using hot plate or spirit lamp. Remember not over 40°C.
k) Add a small drop of warmed glycerine jelly and add the cover slip.
l) Warm the slide for about 5 minutes.
m) Examine the sample under the microscope x400. Identify and count the different types of pollen in each field of view.
n) Seal (if required) when cool as before.

Sedimentation Method.

If you have not got access to a centrifuge take 10g of honey with the added water to make approximately a 30ml solution in the honey jar. First warm honey, especially if crystallised. Stir well with a glass rod. Cover and leave in a cool, draught free place for at least 20 hours.

a) Decant almost all of the liquid leaving about 10ml with the sediment in the jar. Swirl the remainder around in the jar with the glass rod and pour into a test tube.
b) Leave the test tube as before for a further 20 hours.
c) Decant the supernatant liquid and dry the remaining sediment.
d) Make a slide as above.

Recognition of different pollens comes with practise. Dorothy Hodges drawings of pollen grains in her book 'Pollen loads of the Honey Bee', published by IBRA, are a great help in identifying pollens.

6.3 Six slides, made by the Candidate, labelled with the date the slide was made, the scientific name and the approximate size, selected from the following list of pollen grains:

> **forget-me-not, dandelion, rape, lime, sycamore, poached egg plant, crocus, willow, heather, hogweed, rosemary, hawthorn, hazel.**

The table below provides details of the 13 pollen grains from which 6 slides must be made for the examination. The syllabus states "the approximate size" is required; consider this to be plus or minus 10% of the values stated in the table below in the absence of further clarification. It should again be noted that hazel pollen is very constant in size and can be used as a marker.

Common Name	Family	Genus	Species	Pollen size	Shape	Surface (aperture number)
Forget-me-not	*Boraginaceae*	*Myosotis*	*spp.*	c. 6mm	Dumbell	Smooth
Dandelion	*Compositae*	*Taraxacum*	*spp.*	c. 30mm	Irregular	Spines + projections (3)
Rape	*Cruciferae*	*Brassica*	*napus*	c. 35mm	Round	Net (3)
Lime	*Tiliaceae*	*Tilia*	*vulgaris*	c.35mm	Triangular	Pitted (3)
Poached egg plant	*Limnanthaceae*	*Limnanthes*	*douglasii*	c. 30mm	Multi-sided	Granular
Crocus	*Iridaceae*	*Crocus*	*aureus*	c.100mm	Round	Granular
Willow	*Salicaceae*	*Salix*	*caprea*	c. 20mm	Round	Net (3)
Hogweed	*Umbelliferae*	*Heracleum*	*sphondylium*	c. 50mm	Long	Granular (3)
Rosemary	*Labiatae*	*Rosmarinus*	*officinalis*	c. 40mm	Multi-sided	Pitted (6)
Sycamore	*Aceraceae*	*Acer*	*pseudoplatanus*	c. 42mm	Triangular	Striate (3)
Hawthorn	*Roseaceae*	*Crataegus*	*monogyna*	c. 40mm	Triangular	Striate (3)
Hazel	*Corylaceae*	*Corylus*	*avellana*	25mm	Triangular	Smooth (3)
Heather / Ling	*Ericaceae*	*Calluna*	*vulgaris*	c. 45mm	Tetrad	Smooth (3)

The number of apertures to be noted on the pollen grain is given in brackets. Where a number is not given it is because these apertures cannot be easily seen using a microscope of power x400. Precise images of pollen grains can only be seen when using an electron microscope. The above list is a guide to assist the Candidate identifying the pollens required by the syllabus to be made into slides for discussion with the Examiner.

The Scientific order of the sequence of families, genera and species are being continually revised. eg. The family Labiatae is now called Lamiaceae. The major reference for the plants names used here is the “Flora Europaea” 1964-1980.

6.4 How the slides were made and how they should be stored for long term use.

The Candidate needs to make a collection of pollen slides from known species and become familiar with their shapes and irregularities. See the previous section 6.2.

Each slide should be labelled, filed and stored carefully in the horizontal position in partitioned boxes for easy access. The collection should be stored in a cool dry dark cupboard.
The BBKA Headquarters used to keep a library of pollen slides made by Rex Sawyer which could be borrowed for a fee. The monochrome pictures in both Rex Sawyer's books on Pollen and Honey identification are a useful guide when identifying unfamiliar pollens.

6.5 How the size of the pollen on the slide was determined.

The size of the pollen grain can be determined by measurement with an eyepiece graticule or by comparison with pollen grains from the hazel tree. See section 2.9 on the use of an eyepiece graticule and its calibration. Pollen grains from the Hazel tree, *Corylus avellana,* are of a consistent size, 25µm, and can be used as size markers. The pollen grains are held between closely fitting catkin scales, when the pollen is 'ripe' openings form between the scales and any movement of the catkins releases the pollen grains. The pollen is produced in great quantities, the hazel tree is monoecious ie. having separate male and female flowers (uni-sexual flowers) but on the same plant.

Preparation and preservation of *Corylus avellana* pollen.

During late January or February obtain a few small twigs of hazel, with the catkins unopened. Place the open end of the twigs in a jar of water on a sheet of white paper and leave in a warm room by a window. Within a few days pollen grains can be collected from the paper or shaken from the yellow catkins. First place pollen in warm water stained with a few drops of 1% Basic Fuchsin. After about 15 minutes check that the grains have taken up the staining by viewing under the microscope. Remove excess staining and rinse the pollen grains in water several times by shaking in test tube or glass bottle and decanting the excess fluid. Add a few drops of phenol as a preservative and store in a labelled airtight glass container. These pollen grains can then be used as a standard on pollen slides by which one can gauge the size of neighbouring pollens. When making a pollen grain slide add the hazel pollen to the slide using a glass rod, dry the slide and check the position of the pollens before adding the glycerine jelly ie. close proximity of pollens required. It is best to endeavour to position specimens adjacent in the centre of cover slip. Add the warm mountant, cover carefully with cover slip. Warm the slide for about 10 minutes. Check that the slide is satisfactory by viewing under the microscope. If all is well seal the edges of the cover slip with nail varnish to preserve the slide.

6.6 Three pollen slides provided by the Examiner.

We're afraid we can't help you on the bit of the syllabus! We would expect an Examiner to present and quiz you on 3 of the 7 remaining in the table of pollens under section 6.3

6.7 An outline of how microscopic analysis can be used to determine the floral sources and geographic origin of honey samples including the need to take into account the over and under representation of pollen in a multi-floral honey.

The microscopic analysis of honey can be done two ways either by centrifuging or by sedimentation as discussed in section 2.6 (c) pages 52 and 53.

6.8 How the presence of honeydew in a honey sample can be detected by microscopic examination.

We raised a query with the Examination board as follows: "Is this to mean that the sample is 100% honeydew honey or a mixture of honeydew honey and multi-floral honey?"

They responded as follows: "It could be either but more likely to be a mixture of honeydew honey and multi-floral honey as 100% honeydew honey rarely occurs in the UK (although common in Germany)."

Pure honeydew honey contains little pollen from nectiferous plants but does contain pollen from wind pollinated plants such as conifers and grasses. As this pollen blows about in the wind it gets stuck in the sticky honeydew and is subsequently collected by the bees. Therefore a sample from a honey containing both floral sources and honeydew will contain the pollen spectrum associated with the specific flowers together with pollen from wind pollinated plants.

In addition, microscopic analysis of the deposit centrifuged from honey containing a significant quantity of honeydew will also reveals algae, fungal hyphae and spores (particularly sooty moulds which add colour to the honey) characteristic crystals, soot, dust and other debris including hairs etc. from the body of aphids and scale insects (they secrete the honeydew after sucking up the sap from the plant). These tend not to be to be found in honey purely from floral sources.

For further study read:

'The pollen loads of the Honey bee' by Dorothy Hodges published by IBRA,
'Pollen identification for Beekeepers ' and 'Honey Identification' by Rex Sawyer published by University College Cardiff Press,
'An Illustrated Guide to Pollen Analysis' by P. D. Moore and J. A. Webb published by Blackwell,
'A colour guide to pollen loads of the honey bee' by W. Kirk published by IBRA.

* ** *** ** *

7. DISSECTION AND ANATOMY OF THE HONEYBEE

7.1 The Candidate shall provide freshly killed workers, demonstrate ability to embed them in wax during the examination and be able to perform and discuss the abdominal dissection as requested by the Examiner.

The old syllabus required dead bees but now they must be fresh? But how fresh is fresh? We would think it reasonable to have them not older than 2 days and that they should be stored in the refrigerator during this time.

7.1.1 This must be the easiest part of the syllabus because anyone contemplating the examination will have practised the dissection at home a few times. The most difficult part is knowing how many to provide? We would have thought that 3 would more than suffice. The equipment required is as follows:

a) A suitable container for the wax. We use an old tobacco tin about 3½" in diameter and about 1" deep. The press on lid is useful for transporting it to the examination or elsewhere.
b) A piece of wire about $^1/_8$" in diameter and about 10" long bent into an "L" shape with one side ¾" long.
b) Dissecting fluid (Methylated spirit and water).
c) Bunsen burner or small gas torch.
d) Dissecting instruments:
- forceps, two fine pairs and one coarse.
- dissecting needle (s).
- very fine surgical scissors.
- scalpel with spare blades, No.11 are quite suitable.
- small pipette with teat about 4in long overall.

e) Few ounces of clean beeswax.

The tin has first to be filled with beeswax to a level ¼" below the rim and allowed to set by cooling slowly for about 6 hours in order to minimise shrinkage away from the sides of the tin. The level is very important because if the wax level is too low it will make the dissection extremely difficult, or rather that part of the dissection where the scissors have to be used in a horizontal plane.

The next part is to prepare the worker bee before embedding in the wax. The bee for dissection should be as fresh as possible and preferably killed just prior to embedding. Remove the wings and legs by clipping them off with the scissors close to the thorax. Heat the ¾" end of the thick wire and place it in the middle of the dissecting dish, now filled with solidified wax, melting a pool of wax marginally bigger than the bee. While the wax is still molten place the bee in the pool with its dorsal side uppermost; it should be sunk to about halfway in the wax. The coarse forceps are used for this operation and are used to hold the bee until the wax has set. The bent wire can be used to re-heat some of the wax at the sides to pile up a little more wax onto the exoskeleton, if necessary, to ensure that it is firmly held for dissection. Ensure that the bee is in the middle of the dissecting dish; it will make the dissecting work easier, particularly when using the scissors. The whole art of this operation is to use a fresh bee. Avoid overheating the specimen otherwise it will make the dissection of the internal viscera difficult.

The last part is to fill the dissecting dish with dissecting fluid to nearly the top of the dish (or tobacco tin) just enough to cover the bee. Water is unsuitable because it will not wet the hairy parts of the bee (a problem of surface tension) so alcohol is required in the form of methylated spirits. A mixture of 60% water and 40% meths by volume is satisfactory. It is a good idea to have a bottle full of this mixture available for the examination. Ensure that it is airtight and well labelled. The alcohol acts as a partial fixative preventing decomposition as well as acting as a wetting agent minimising the many minute air bubbles which will form in the internal viscera as well as in the plumose hairs on the outside of the skeleton.

7.1.2 The step by step process in dissecting the alimentary tract.

The best book on the subject is "Anatomy and Dissection of the Honeybee" by H.A.Dade first published in 1962. It is essential reading for this examination and describes the dissection of the alimentary tract with diagrams to perfection.

The Examiners will require the Candidate to describe the internal viscera after the roof of the abdomen has been removed including the heart and associated ostia in the roof itself. The Candidate will be expected to comment on the state of the bee by observation of the honey sac and rectum. The whole of the alimentary tract should be 'teased' out and laid to one side to reveal the underlying organs, eg. workers' ovaries which are not too easy find.

The power of the dissecting microscope is important and is related to the dexterity of the operator. The best magnification is one which the individual feels 'comfortable' with; this is likely to be within the range of 5 to 20. It is essential to experiment to decide what is best for you.

Finally, when you have finished clean and dry all the instruments and protect the points of them all by pushing them into spare scraps of polystyrene.

7.2 The Candidate shall make and provide for discussion four labelled anatomical slides with one slide made from each of the following four lists. At least three of these slides should be prepared as permanent hard mounts.

List A	List B	List C	List D
Front, middle and hind leg of a worker	Mouthparts of a worker (displayed)	Fore and hind wings of a worker or drone	Sperm
Comparative slide of a hind leg from a drone, queen & worker	Comparative slide of mandibles of adrone, queen and worker	Antennae of both a worker and drone	Two Varroa mites (one mounted dorsally and the other ventrally)
Everted endothallus of drone genitalia	Sting of a worker	Portion of trachea	Hypopharyngeal gland

7.2.1 General.

"Microscopy on a shoestring for Beekeepers and Naturalists" by Owen Meyer published by Northern Bee Books in 1984 contains 4 very pertinent chapters relevant to this part of the syllabus and we regard it as very useful reading for this examination. However, some of the sizes quoted by Meyer for some disease pathogens are, in our opinion, suspect. Additionally, "Anatomy and Dissection of the Honeybee" by H.A.Dade is the de facto text book on the dissection of the honeybee; old (first published 1962) but very pertinent.

In the table of possible slides, list B includes the mouthparts of a worker displayed; The Examination Board state that 'displayed' means "arranged to show all/most of the parts of a dissection etc. in one view". In the case of mouthparts, the parts are arranged on the slide sufficiently spread out to see them all.

Before considering the slides to be made in this section of the syllabus it would be desirable to discuss a few terms and some of the materials used in their preparation. An understanding of the terminology will assist in the successful preparation of the slides for long term use. The basic objective in making a slide of any animal part is to present it, as near as possible, in its original form. As soon as death occurs in any animal, and the bee is no exception, the cells cease to work and decomposition commences more or less quickly dependent on the ambient temperature. The mechanism is twofold, one by autolysis which is the destruction of the body cells by their own enzymes and two by attack from any bacteria which are generally present.

Fixation means the prevention of decomposition, in this case, by chemical methods and not by low temperatures. Alcohol is not a particularly good fixative as it penetrates the tissues only slowly and makes the tissues brittle. Formaldehyde (aldehyde of formic acid, HCHO) has the advantage of quick penetration and toughening the tissues at the same time. Acetic acid is the third fixative which has fast penetration and exhibits no hardening of the tissues. It is for these reasons that most fixatives contain formaldehyde or formalin (an aqueous solution of formaldehyde) and / or acetic acid. There are many fixatives and the more expert the microscopist the more they favour one or another for a variety of reasons. Two will be adequate for our purposes as follows:

Formalin - alcohol	Alcohol (70%)	94% by volume
	Formalin (40%)	6% by volume
Formo-acetic-alcohol	Alcohol (70%)	88% by volume
	Acetic acid (100%)	7% by volume
	Formalin (40%)	5% by volume

Staining means just what is says but it should be noted that there is positive and negative staining. Positive staining stains the specimen while negative staining stains the background or mountant containing the specimen. There are many stains available for a variety of purposes and the skilful microscopists have their own preferences. For our purposes simplicity is the key to success for the simple microscopy we are considering and we list 4 common stains which will be more than adequate for the syllabus as follows:

Fuchsin (Aniline red) - A little goes a long way! 2 or 3 crystals dissolved in alcohol and added to 200 ml of water will last a life time. It is used in glycerine jelly as a prepared mountant. Very widely used.
Nigrosin black - Used as a negative stain particularly for slides of bacteria (eg. AFB, EFB) and protozoa (Nosema apis). Again a little goes a long way. Meyer uses Indian ink which he claims gives better results. We have never used it; still using up our stock of nigrosin!
Gentian violet - used for staining cell nuclei.
Methyl blue – Ditto.

Differentiation is just the opposite of staining for some very obscure reason, unknown to us at least. It is a method of removing excess stain to improve the contrast between the stained parts and their surrounds. Again keeping things simple, two de-staining agents will be adequate for our purposes as follows:

Milton	Used with fuchsin, nigrosin and methyl blue
Clove oil	Used with gentian violet

Dehydration. It is necessary to remove any water from the specimen because it will not mix with the mountants used in the final preparation of the slide. Failure to remove any water will result in minute bubbles in the slide. Alcohol (100%) is very hygroscopic and, because of this, it tends to distort the specimen as the water is drawn out quickly. Acetone is cheaper and more readily available as well as proprietary brands of dehydration fluids. Again, as in all microscopy, it is a matter of taking your pick as competence and experience is gained.

Clearing is the removal of alcohol or dehydration fluid. Failure to do this will result in a milky film being formed. This is readily demonstrated by putting a little acetone onto a damp surface; it turns milky almost immediately. Two popular clearing agents are **xylene** and **cedar oil**.

Mounting is the placing of the specimen in a suitable medium (called the mountant) and covering with a cover slip avoiding the introduction of air bubbles. The two most widely used mountants are **glycerine jelly** and **Canada balsam**; however, again there are proprietary substitutes available from specialist suppliers. It is important to note the following points in respect of both these mountants:

a) Like most materials they will discolour if exposed for long periods in bright light. It is important that finished slides are stored in a dark cool dry place.
b) The minimum of mountant required should be used to prevent it oozing out from under the cover slip when it is in position; it is very difficult to clean off before ringing.
c) Canada balsam takes a long time to dry. Slides should be left for 48 hours before ringing.
d) Glycerine jelly is available already coloured with fuchsin which is very useful for making pollen slides as the staining and mounting can be undertaken in one operation.

Ringing means sealing the edge of the cover slip to the actual glass slide. We assume that the name originated from the circular cover slips which were ringed by placing them on a rotating table, the centre of the slip being in line with the axis of rotation, and brushing the edge carefully with the sealant. We have used **nail varnish** for ringing, it comes in handy sized bottles with the right type of brush for the job on hand (no pun intended). If acetone is used as a dehydration fluid then it can also be used as a solvent for the nail varnish.

A further point worthy of note is about bleaching of specimens. If they are to be finally viewed with transmitted light in a compound microscope, darker specimens may require to be bleached to allow the passage of light. In the list, from which to select a sample slide for the examination, the antennae will be improved by bleaching. Domestic bleach is satisfactory for the purpose cleaned off with water to stop the bleaching process when the specimen is considered to be light enough. All chitinous parts may require bleaching and the slide of the mouthparts could be also usefully bleached.

One final point before actually making the slides it is important that the 1in x 3in glass slides are meticulously clean and free of all grease. Most books suggest hand washing with detergents and air drying to avoid fluff from a drying cloth. We believe this to be tedious and not entirely satisfactory. If a holder is made for the glass slides to go into the dish washer with the slides in a vertical plane, they will be washed better than any hand washing process and come out sparkling and completely grease free.

7.2.2 Making the slides.

When a Candidate presents his 4 slides at the examination the Examiner will ask questions about the slide which can include:

> How the specimen was obtained, how it was killed and details of how the dissection was performed together with what chemicals/materials were used in their preparation.

There is a lot of detail here, so be prepared.

There is a wealth of information available about making slides (some sources have been indicated) and only brief notes follow about each slide providing information that cannot be found elsewhere with tips which, in our opinion, will make the practicalities easier.

7.2.2.A1 Front, middle and hind legs of a worker.

The legs are quite large both in length and thickness so the final slide will require a relatively large amount of mountant. Should the slide show the inside or the outside view of the three legs? Preferably two slides would be better, one showing the outside of the legs and the other showing the inside; they could all come from the same bee. The syllabus is not clear on this point. Cavity slides are essential.

The legs contain tendons, muscles and haemolymph so they will require this to be dissolved after bleaching the chitinous exterior. A 10 to 15% solution of caustic potash (potassium hydroxide) for about 3 days should complete the maceration which should be followed by washing, dehydration, etc.

Mounting the legs is not easy arranging them in an orderly fashion parallel one with another. Super glue comes to the rescue here but only the tiniest amount to hold the legs in place. Try with some unprepared legs first on an old bit of glass to get the feel of it. Finally cover with warm mountant and cover slip in a cavity slide left to set before ringing. There is nothing magic in the preparation of such a slide, it is all a matter of practice and experience.

7.2.2.A2 Hind leg from a drone, queen and worker.

This slide is much the same as A1.

7.2.2.A3 Everted endothallus of drone genitalia.

See Dade page 122 and Plate 15.

7.2.2.B1 Mouthparts of a worker (displayed).

After discussion the Authors are of the opinion that the syllabus is imprecise and should have stated clearly that it was both the proboscis and the mandibles that are required.

Both Dade and Meyer use the same method for preparing the proboscis part of this slide which involves removing the head of a worker honeybee and pinning it face down with an acarine pin with the needles either side of the foramen. We have found modern technology better by using super glue to stick the head face down on an old slide. It makes the dissection easier without the acarine needle getting in the way. Try both methods for yourself and find out which you are most comfortable with. We recommend that you follow Dade and articulate the proboscis with a needle before undertaking the dissection described in Chapter 10 section 1 on page 106. The slide preparation is described in Appendix II on page 140.

7.2.2.B2 Mandibles of a drone, queen and worker.

If reference is made to Dade's Plate 3 showing the mandibles it will be clear that the syllabus doesn't state whether the inside surface or the outside surface should be shown in the prepared slide. Take your pick but don't muddle them up.

7.2.2.B3 Sting of a worker or queen.

There is a very good description of how to deal with this slide in Meyer's book which is worth referring to. It appears to be identical with Dade who describes also another method mounting the sting in glycerine and retaining all the muscles and glands rather than macerating them and then flattening the plates by removing the proctiger. Dade's second method produces a very thick section requiring a cavity slide. Note that Dade made his own cavity slides using the ringing turntable and gold size followed by double ringing with nail varnish and then with Araldite.

7.2.2.C1 Fore and hind wings of worker or drone.

This must be the easiest slide of the whole selection available for the examination as they require no preparation. However, note that Dade states that they should be dehydrated in Cellosolve an alternative to alcohol; it is not clear to us why he says this. Glycerine or Canada balsam is a suitable mountant.

Because of the simplicity of this slide an Examiner may wish to probe about its probable uses and asking questions about the venation, cubital index, hamuli, etc.

7.2.2.C2 Antennae of both worker and drone.

Endeavour to remove both antennae without damage to the posterior end of the scape. Both will require partial bleaching and maceration before mounting, Meyer gives a good description of the requirements in chapter 7 on page 87.

7.2.2.C3 Portion of a trachea.

The easiest trachea to dissect is the first thoracic trachea revealed when examining for acarine. A trachea with a mite in it is even more interesting and the slide then shows two features.

Reference to Dade and Meyer, they indicate two different methods of preparation. Dade fixes in alcohol, stains, rinses, dehydrates, clears and then mounts in Canada balsam, whereas Meyer soaks in glycerine and then mounts straight into glycerine jelly. This demonstrates nicely that, as with many things, there are more ways than one to skin a cat.

We have just examined one of our own slides made in 1990 which was mounted straight from the bee into warmed glycerine jelly showing trachea and *Acarapis woodi* eggs, nymphs, adults and the faecal debris in the trachea. No deterioration of the slide could be detected. We suggest that you try a variety of methods.

7.2.2.D1 Spermatozoa.

Each spermatozoon has a head 10µm long and 0.5µm diameter with a tail 250µm long. In order to see these it will be necessary to stain them. The tails are normally doubled and coiled two or three times when taken from the drone or from the queen. It is best if these wound up tails can be unravelled before killing and staining.

Spermatozoa may be obtained from the seminal vesicles of a mature drone or from the spermatheca of a recently killed fertile queen. A sample should be put on a watch glass with a drop of warm water (100º F / 38ºC) and kept warm for a few minutes for the sperm to become active and the tails to uncoil.

a) With a clean glass rod put part of the specimen on a slide and spread it out to a size smaller than the cover slip and then allow it to dry completely without extra heat.
b) Add a drop of stain (Dade recommends carbol fuchsin - diluted) and warm the slide gently. Do not allow to dry. The staining will take about 2 hour. If the stain starts drying add more.
c) Wash very gently with water under a trickling tap to remove stain. Leave the tap trickling.
d) Put on one drop of alcohol (pipette) and immediately wash off under the aforesaid trickling tap and allow the specimen to become completely dry.
e) Apply mountant and warm cover slip. Keep warm to allow the cover to settle down .
f) Ring with nail varnish.

7.2.2.D2 Two varroa, one mounted dorsally and the other mounted ventrally.

This is such a simple slide to make it requires no explanation - bingo again for super glue.

7.2.2.D3 Hypopharyngeal gland.

The first operation is to dissect the glands. There are two ways to do this either by removal of the dorsal part of the head exposing the glands from above or by removing the anterior portion of the head ('the face'). The latter seems to be the most preferable. The dissection should be done under water and if the head is removed and glued to an old piece of slide with super glue the glass can be fixed into the wax in the dissecting dish. Then follow Dade Chapter 12, section 2 on page 119 to expose the glands.

Dade suggests fixing, staining and mounting all in one operation which is fine if you have some lacto-phenol mountant available to which should be added a little blue dye. Otherwise proceed to undertake each task separately it is very easy. Note that only a small portion of the gland is required for the slide. It is better to be able to view a single layer of acini rather than a thick clump. The advantage of doing the various operations separately is that glycerine jelly or Canada balsam can be used as the mountant making the ringing process easier and surer.

7.3 Some final comments.

a) Try to include a class at your branch and county shows for two or three set slides each year in order to improve the quality of our microscopists. We know of one or two associations that do this already but it is not as widespread as it should be. The National Honey Show could usefully set a lead in this direction.
b) County associations should be encouraged to purchase a pair of suitable microscopes and a stage micrometer for calibrating the compound microscope. These can be lent to association branches or those preparing for BBKA examinations.
c) Adopt a disciplined approach and keep a log book of your slides, it easy to forget in a few years time what materials were used in the preparation of a particular slide.
d) The materials required are generally in much smaller quantities than packaged for purchase. Sharing them and the cost between interested parties is a sensible approach. For example, we purchased the smallest pot of stained glycerine jelly we could find in the 1980s and we still have more than 90% left after making a few hundred slides.
e) Do not be afraid to experiment with different methods. It will be clear from what has been said that there are many options open to the lateral thinker.
d) See appendix 3 for details of where to purchase microscopes and materials for slide making. A search on E-bay has just revealed a bargain in compound microscopes!

7.4 Further useful reading.

"Beekeeping under the microscope by T. Maxwell" - 3 monographs Part 1, 2 and 3 published by the Central Association.
"Setting up the microscope by G. W. White" - an 8 page monograph published by The Quekett Microscopical Club.

* ** *** ** *

8.0 HEALTH AND SAFETY

The Candidate shall discuss with the Examiner:

The whole of this section 8 of the syllabus has come about as a result of EU legislation which has been bureaucratised by the Health and Safety Executive in this country. It is mostly for laboratory use but can be regarded as a useful guide for working at home as most Candidates will be doing. All it eventually boils down to is plain common sense. Unfortunately the HSE carry a lot of clout and deviation from the legislation can involve some pretty hefty fines. So, until the UK manages to leave the EU, do take heed. This section only involves a discussion with the Examiner.

8.1 The potential hazards of working with chemicals, naked flames, microscopes, electrical devices & dissecting instruments.

Consider each in turn:

Chemicals. All substances can be considered dangerous, some more than others. The chemicals used in microscopy range from very low risk (distilled water) to very high risk (acetic acid). They can cause damage to the eyes, skin and damage to lung tissue if fumes are inhaled. Some are extremely flammable and some can cause an explosion. See Appendices 4 and 5 for more information.

Naked flames. Can cause burns, fire and explosion. Take care and keep flammable chemicals well away from Bunsen burners or gas torches used for processes requiring their use.

Microscopes. We believe that there are three points to note with respect to microscopes:- a) Ensure that they are fastened securely before moving them, b) Ensure that they are used well away from sunlight falling onto the condenser lens/mirror and c) be aware of the correct fluids for cleaning objective lenses.

Electrical devices. Stating the obvious, faulty equipment can cause electric shock.

Dissecting instruments. Most are sharp and pointed and can cause cuts and wounds and possible infections.

8.2 The need to undertake a risk assessment before commencing any activity of a practical nature.

We approached the Examination Board about this section of the syllabus and they replied as follows:-

> *"The health and safety aspect was requested from several people. The candidate is expected to work safely with respect to themselves and others appreciating and avoiding any potential hazards involved.*

This is assessed partly by asking the candidates specific questions during the assessment.
Typical questions could include:

Why should you do a risk assessment before carrying out a procedure ?
Which substances should be kept away from a naked flame ? etc.)
Which chemicals used to produce your anatomy slides are potentially hazardous & what are the main hazards associated with these chemicals.?
What precautions do you take when handling hazardous chemicals?
Identify potential hazards of using a naked flame to heat your embedding tool?
How do you safely dispose of waste chemicals (or dead bees, broken glass slides, cover-slips, scalpel blades etc.)?

In addition assessors monitor the candidates during the whole assessment period to ensure they worked safely with respect to themselves and others in the room and to leave comment where consideration of safety by a candidate was lacking in any way."

In the above read Examiner(s) for assessor(s). Note that the Board say "...assessed partly...." we therefore assume that the other part of their assessment is by observing the Candidate while undertaking a dissection.

Now we come to the tricky bit. 'Elfansafety' require a written risk assessment to be completed before any activity takes place. The examination is itself a practical activity (a bee has to be embedded in beeswax and then dissected). Who undertakes the written risk assessment? Is it the Examiner or is it the Candidate? The Examination Board didn't reply to this question. In our opinion, the responsibility for doing a written risk assessment should be the responsibility of the Examiner rather like the teacher in the chemistry lab, not the pupils.

Finally, see the end of Appendix 5 "safety in the Laboratory" for a simple straightforward risk assessment form.

8.3 The need to wear protective clothing, chemical resistant gloves and goggles when handling hazardous chemicals and to always work safely with respect to themselves and others.

Please see Appendix 5 for further details of this new syllabus item.

8.4 The safe disposal of hazardous waste e.g chemicals, broken glass slides, scalpel blades and remains of bees.

This is a new item in the syllabus and involves a plethora of EU regulations. Because of the tens of thousands of regulations which have emanated from Brussels and because it is so easy for the man in the street to inadvertently transgress, we thought it prudent to approach our local district council on this matter. They are responsible for waste collection and disposal. The following e-mail was sent to them:-

I would be most obliged if you could advise me on the approved way of disposing of the following items used in microscopy work:-
a) Chemicals: Propanol, Xylene, Acetic acid, Sodium hydroxide, Ethyl acetate, Glycerin jelly, Fuchsin and Canada balsam all in small quantities of not greater than 100ml.
b) Broken glass slides.
c) Scalpel blades.
This is part of a syllabus for an examination in microscopy and I would like to be up to date with the latest approved methods complying with the law.

who ducked the question and in turn sent it Devon County Council who have replied as follows:-

I would advise you to contact the Environment Agency for up to date info on the handling and disposal of the items concerned.
We do not consider them to be household waste and so could not accept them at our facilities, however specialist hazardous waste companies are there to deal with these waste types.

We then sent the same query to the Environment Agency who replied as follows:-

It is the responsibility of the waste producer to classify their waste and to locate an appropriate operator for its disposal. We have guidance available on our website to help you classify your waste and determine whether the waste is classed as hazardous or not. Please see the link to the European Waste Catalogue (EWC) on the following webpage for details:

Coding and Classifying Waste:
http://www.environment-agency.gov.uk/business/topics/waste/32140.aspx

The European Waste Catalogue (EWC) is a classification system for waste materials. It categorises wastes based on a combination of what they are and the process or activity which produced them. Every Waste is identified by a six digit code, often referred to as a "EWC Code". This code also identifies whether the waste is hazardous or not. There is a legal requirement to use this code on all duty of care transfer notes and consignment notes. Again, we are unable to say which codes you would need to use for the waste described in your email. It would be the waste producer's responsibility to locate the correct code.

Further information on classifying and assessing waste is also available in the Technical Guidance WM2. A copy of which can be downloaded from the following link:

Hazardous Waste:
http://www.environment-agency.gov.uk/business/topics/waste/32180.aspx

When contacting a company to collect waste from you, you have a Duty of Care to ensure that they are legally allowed to take waste from you. Anyone wishing to collect waste will need a waste carrier/dealer registration or a waste carrier/broker/dealer registration. Information on waste carriers is available on our website at:

Waste Carriers/Brokers and Dealers:
http://www.environment-agency.gov.uk/busintess/sectors/wastecarriers.aspx

For information on locating waste carriers, please refer to Alexander's email.

You can check if a registration is valid by checking the registration on our online Public Register. A link to the register is shown below:

Public Registers:
http://www.environment-agency.gov.uk/research/library/publicregisters/default.aspx

You may also find the following links helpful:

Your waste responsibilities:
http://www.environment-agency.gov.uk/business/topics/waste/121280.aspx

Duty of Care:
http://www.environment-agency.gov.uk/business/topics/waste/40047.aspx

We trust that this information will be useful.

So not even the ultimate authority in the UK is prepared to advise how to dispose of very small quantities of waste material used for microscopy in the home. After checking all the web sites above, it would appear that an ordinary householder has to make all the arrangements to retain the services of a host of different contractors (one for each chemical), make payment to them and then, would you believe it, shoulder the responsibility of ensuring that they are correctly licensed to dispose of very small amounts of waste. This could be the official answer to this part of the syllabus, albeit not a very pragmatic one. NB. The email from Alexander referred to above has not been reproduced as it adds nothing useful to the discussion.

The Authors have been doing their own microscopy for many years and any sharp tools have had the edge ground off them before putting them in the bin. Any broken glass (eg. slides or cover slips) are well wrapped in plenty of newspaper and again put in the bin. Any waste chemicals are flushed down the toilet together with any waste bee samples. We have not checked whether we have breached any EU instigated laws but we believe that we have acted sensibly and with responsibility.

Many drains in this country are unblocked with very hazardous acids and in quantities far greater than are envisaged here. The only sewerage detection that we know about is for chemicals used by terrorists for bomb making and this is in Sweden and, at the time of writing, is in its experimental stage. Also at the time of writing, we hear that the EU are drafting regulations to enforce a standard flush for all toilets throughout the Federal States of Europe! It beggars belief.

So, dear reader/Candidate, you will have to make your own mind up about how to answer questions on this topic should they arise in your examination.

* ** *** ** *

APPENDIX 1

DISEASE SAMPLING

Proportion of diseased bees in the sample N = f_o (observed)
Proportion of diseased bees in the whole population = f_a (actual)
Standard deviation of sample proportions,

$$\sigma = \sqrt{\frac{f_a(1 - f_a)}{N}}$$

Consider $\frac{\Delta f}{\sigma} = R$ where $\Delta f = (f_a - f_o)$. [R = number of standard deviations the observed population differs from the true population.]

Therefore, $$\frac{f_a - f_o}{\sqrt{\frac{f_a(1 - f_a)}{N}}} = R$$ and thus, $$f_a^2 - 2f_af_o + f_o^2 = \frac{R^2f_a}{N} - \frac{R^2f_a^2}{N}$$

Therefore, $$f_a^2(1 + \frac{R^2}{N}) - f_a(2f_o + \frac{R^2}{N}) + f_o^2 = 0$$

Note the solution to the above quadratic equation, of the form $ax^2 + bx + c = 0$ is:

$$x = \frac{-b \pm \sqrt{b^2 - 4ac}}{2a}$$ hence, $$f_a = \frac{(2f_o + \frac{R^2}{N}) \pm \sqrt{4f_o^2 + 4f_o\frac{R^2}{N} + \frac{R^4}{N^2} - 4f_o^2 - 4f_o^2\frac{R^2}{N}}}{2(1 + \frac{R^2}{N})}$$

Using the larger of the two roots: $$f_a = \frac{2f_o + \frac{R^2}{N} + \sqrt{4f_o\frac{R^2}{N}(1 - f_o) + \frac{R^4}{N^2}}}{2(1 + \frac{R^2}{N})}$$

Now $f_o = \frac{1}{N}$ for only one diseased bee found in the sample

therefore, $$f_a = \frac{\frac{2 + R^2}{N} + \sqrt{\frac{4R^2}{N^2} - \frac{4R^2}{N^3} + \frac{R^4}{N^2}}}{2(1 + \frac{R^2}{N})}$$ (as shown on the graph overleaf)

R is derived from the area under a normal curve $$\int_{-\infty}^{z} \frac{1}{\sqrt{2\pi}} e^{-\frac{1}{2}z^2} dz$$

eg. R = 2.33 for 99.01% probability. See any book of statistical tables.

Doing the calculation to plot the curves is unduly tedious by manual computation and they were computed by computer program. Listing the the results for drawing the graphs serves no useful purpose.

We have shown the derivation of the formula for two reasons, a) we believe that the probabilities provide by Deans et al are at variance with the above and b) when we included the graphs in the Senior Part 1 Study Notes we had numerous enquiries seeking the derivation of the formula. Unfortunately, we have not been able to track down the detailed working of the formulae used by other authors to check their validity.

Graph of percentage colony infection if one bee is found in the sample.

The graph, on the following page, is straightforward to use and it is derived as shown previously on the assumption that one bee in the sample is diseased. If more than one bee in the sample is found to be diseased, then the colony infection can be calculated as follows:

> If the sample size = x
> and the number of diseased bees = y, where y < x
> Then N becomes x ÷ y and the graphs used in the normal manner.

eg. 3 diseased bees in 30, then N = 10 giving a 99% possibility that the colony has 47.5% of its population infected (from the graph).

Branch microscopists should be encouraged to supply such information on the report form returned to the beekeeper after diagnosis has been completed. A typical form, which has been in use in the Plymouth Branch for sometime is shown in Appendix 2. We are of the opinion that the graph is easier to use than a table of figures which are generally based on 1 diseased bee in 30 and invariably the sample never equals exactly 30 bees. Our graphs allow for any number of bees (match box full) in the sample which may be sent the microscopist.

* ** *** ** *

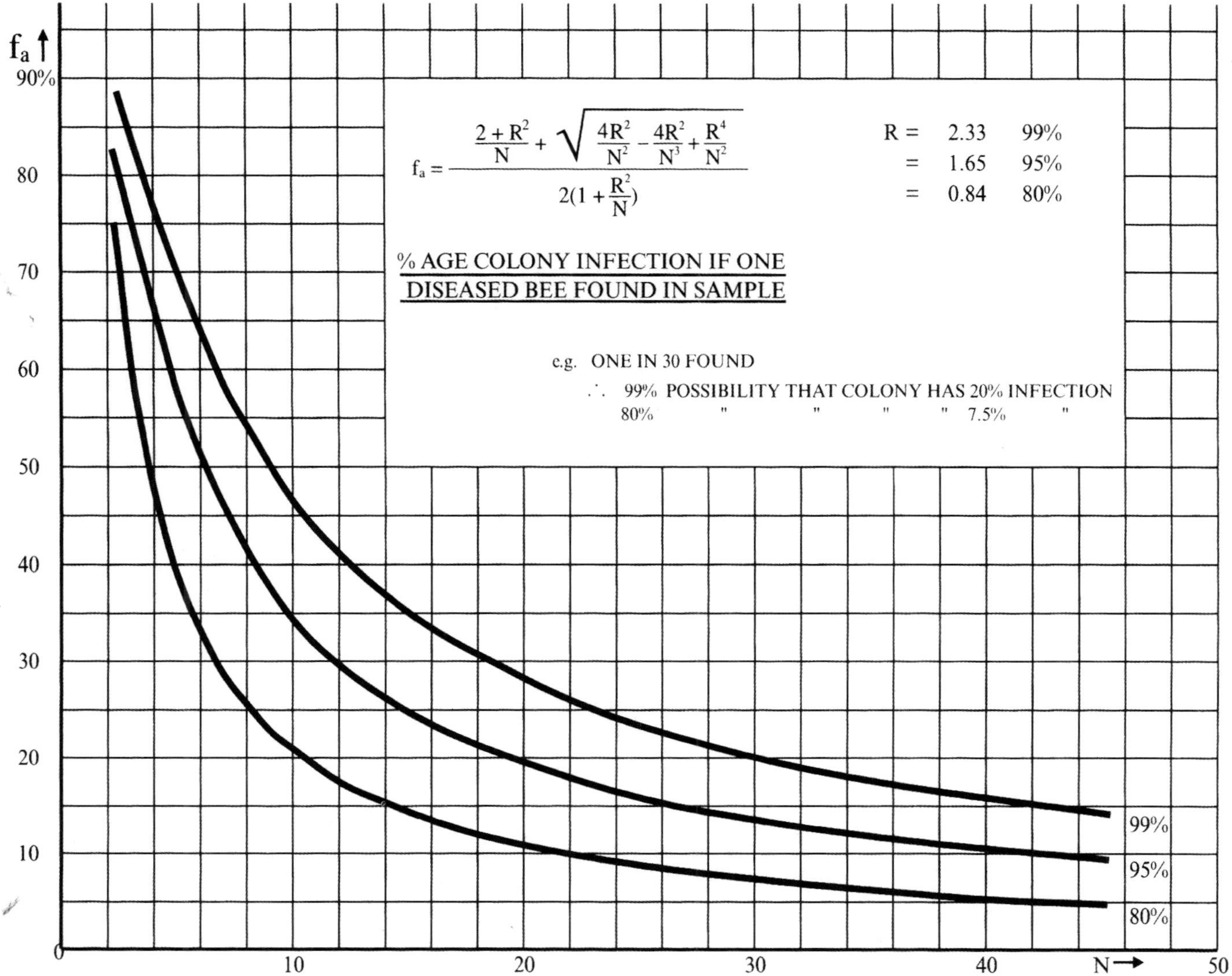

f_a ↑
90%
80
70
60
50
40
30
20
10
0
10
20
30
40
N →
50
$f_a = \dfrac{\dfrac{2 + R^2}{N} + \sqrt{\dfrac{4R^2}{N^2} - \dfrac{4R^2}{N^3} + \dfrac{R^4}{N^2}}}{2(1 + \dfrac{R^2}{N})}$
R = 2.33 99%
= 1.65 95%
= 0.84 80%
% AGE COLONY INFECTION IF ONE DISEASED BEE FOUND IN SAMPLE
e.g. ONE IN 30 FOUND
∴ 99% POSSIBILITY THAT COLONY HAS 20% INFECTION
80% " " " " 7.5% "
99%
95%
80%

APPENDIX 2

Ref. No.. ....

MICROSCOPIST REPORT - ADULT BEE DISEASES
(PLYMOUTH BRANCH of DEVON BEEKEEPERS' ASSOCIATION)

SAMPLE RECEIVED FROM: ....

Address: ....

....

....

Date sample (s) received: ....

Condition of sample: good / adequate / poor

RESULTS OF ANALYSIS:

SAMPLE		1	2	3	4	5
	No. of bees in the sample					
ACARINE	Percentage of bees infected					
NOSEMA	Light					
NOSEMA	Medium					
NOSEMA	Heavy					
AMOEBA						

Acarine: The percentage of bees infested in the colony shown above is based on a 99% probability.
Nosema: When detected is based on the following classification:

Light 0 to 5 spores (using 400 magnification, objective = 40)
Medium 6 to 15 spores (ditto)
Heavy >15 spores (ditto)

RECOMMENDATIONS FOR TREATMENT:

Acarine: Spring - nil: Autumn – No specific treatment is currently available. Treatment for Varroosis, using Bayvarol or Apistan, has been found to be effective for treating Acarine.

Nosema: Spring (light and medium) - FUMIDIL B; see notes f) and g).
(heavy) - FUMIDIL B + Bailey frame change and re-queen as soon as possible.
Autumn - FUMIDIL B with winter feed for all levels of infection; see noted f) and g).

PTO.

OTHER OBSERVATIONS:

Signed: ....(Branch Microscopist)

Date: ....

Notes:

a) There is no known treatment for Amoeba.

b) Contact your Branch Secretary for Fumidil B supplies.

c) For microscopes with objectives other than x40 divide the number of spores by the square of the new objective and then multiply by the square of the old objective (1600) to obtain the new spore number; eg. for x60 instead of x40

the new spore number = (old spore number x1600) ÷ 3600 = 0.4 x old spore number

Thus	light	would be	0 to 2 spores
		medium	3 to 6 spores
		heavy	>6 spores.

d) The classification for nosema has been recommended by JD & BD Yates, it is empirical and not based on theory. Similarly, the recommended treatment is based on their experience managing their own apiaries.

e) Members of the branch are permitted to send a maximum of 5 samples both in the spring and in the autumn for analysis unless special arrangements are made with the microscopist.

f) It should be noted that Fumidil B ceased to be licensed after 31st December 2011, any stocks remaining after this date could be used for an extended period of 6 months.

g) There is some untested evidence that Thymol in sugar syrup may be beneficial for light infections of nosema.

* ** *** ** *

APPENDIX 3. Suppliers of equipment and materials.

Now where to get all the materials for undertaking the preparation of the slides?

Importers and suppliers of optical instruments.

Brunel Microscopes Ltd., Unit 2, Vincients Road, Bumpers Farms Industrial Estate, Chippenham, Wiltshire SN14 6NQ (Telephone: 01249 462655)

Lakeland Microscopes, Hollybank, Windemere Road, Grange-over-Sands, Cumbria LA11 6LB (Telephone: 01539 534737)

Materials for slide making.

Chemicals, laboratory equipment and biological supplies.

Contact Timstar Laboratory Suppliers Ltd., Unit S1, Herald Park Industrial Estate, Herald Drive, Crewe, Cheshire CW1 1EA Telephone: 01270 250459, Fax 01270 250601 who have a very comprehensive range of everything required for this examination.

Other useful contacts.

Manchester Microscopical and Natural History Society, Hon. Sec., 2 Borwick Close, Warton, Carnforth, Lancashire LA5 9QY (Telephone: 01524 732533). This society publishes "First Steps into a Secret World" (BBNO, £3.99) which is a useful publication for beginners and has an extensive glossary.

* ** *** ** *

APPENDIX 4. The CLP Regulation.

The GHS (Global Harmonised System) system of labelling chemicals, used for many years throughout the world has been superseded from the old familiar CHIP (Classification Hazard Information & Packaging) system to a new one known as CLP (Classification Labelling & Packaging). The new system is used only in the EU while the rest of the world retain the GHS system; not exactly progress in our opinion. The following was taken from the Health and Safety web site www.hse.gov.uk :

The CLP Regulation

European Regulation (EC) No 1272/2008 on classification, labelling and packaging of substances and mixtures came into force on 20 January 2009 in all EU Member States, including the UK. It is known by its abbreviated form, 'the CLP Regulation' or just plain 'CLP'.

The CLP Regulation adopts the United Nations' Globally Harmonised System on the classification and labelling of chemicals (GHS) across all European Union countries, including the UK.

As GHS is a voluntary agreement rather than a law, it has to be adopted through a suitable national or regional legal mechanism to ensure it becomes legally binding. That's what the CLP Regulation does.

As GHS was heavily influenced by the old EU system, the CLP Regulation is very similar in many ways. The duties on suppliers are broadly the same: classification, labelling and packaging.

The existing legislation on classification, labelling and packaging has been agreed at European Union level and, from 2015, will be directly applied on all EU member states, including the UK.

The rules they have to follow when they are classifying will change though, and a new set of hazard pictograms (quite similar to the old ones) are used:

Old CHIP symbols

These symbols used to help us to know that the chemicals we are using might be explosive, oxidising, highly or extremely flammable, (very) toxic, harmful, irritant, corrosive, or dangerous for the environment. One or more might appear on a single chemical. They were known as GHS (Globally Harmonised System) pictograms. (Colour reproductions of these symbols can be seen on page 46 above.)

These symbols are being replaced by others because the law on chemical classification and labelling has been changed by the EU.
The new symbols, called pictograms, show similar images just a slightly different shape and colour.

New CLP symbols

You'll see that the harmful symbol is missing. This has been replaced by the exclamation mark pictogram:

This pictogram will refer to less serious health hazards such as skin irritants / sensitisation.

A couple of new pictograms have also been introduced:

This pictogram reflects serious longer term health hazards such as carcinogenicity and respiratory sensitisation.

This pictogram means "Contains gas under pressure". (Colour reproductions of these symbols can be seen on page 46 above.)

The new look labels

These new pictograms will be accompanied on a chemical label with various other items of information which amplifies their meaning with **Hazard Statements**, and handling instructions now known as **Precautionary Statements**.

Other information should include the name and address of the supplier, the name and product identifier (usually an EC No.), the quantity of the product in Grams or Kilograms, or if a liquid, in Millilitres or Litres. There must also be a 'Signal Word' of either '**DANGER**' or '**WARNING**' for less hazardous materials.

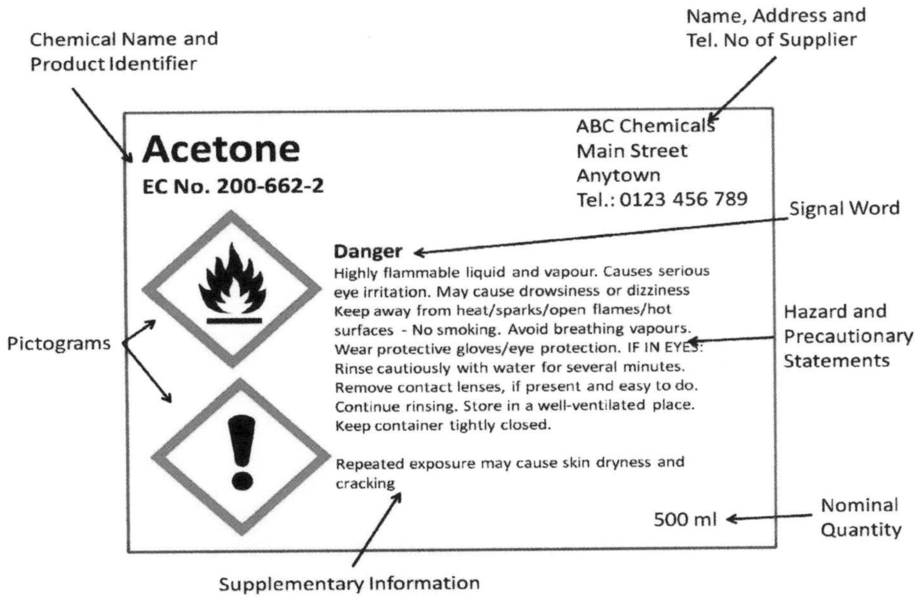

* ** *** ** *

APPENDIX 5. Safety in the Laboratory.

The following was prepared by John Chandler of the Harrogate & Ripon BKA for his own classes that he conducts for this examination.

Safety in the Laboratory.

Some essential guidance for those studying for the BBKA Microscopy module.

These notes are intended as a basic guide to assist Microscopists in handling unfamiliar chemicals. It also gives some advice on how to prevent the spread of infection whilst investigating diseased bees and some tips about the care and safety of equipment.

John Chandler
Harrogate and Ripon BKA.

June 2009 (revised 2012)

Index

Safety in the Laboratory.

A summary of the general rules that apply to laboratories or to where you work with chemicals at home:

1. Do not eat, drink or smoke where you are using chemicals or where there may be biologically hazardous materials (diseased bees for instance).
2. Wear personal protective equipment (disposable gloves and anti splash goggles -with side protection - when handling chemicals or bio-hazardous material). Lab coats protect clothes and skin.
   - See advice on types of disposable gloves on page 9.
3. Label all bottles and containers with appropriate warnings where necessary. (See page 8 for information on these.)
4. Work in a well ventilated area (avoid the cupboard under the stairs!) when using volatile chemicals such as propanol, acetic acid and xylene.
5. Store chemicals in dark, cool and (if possible) ventilated areas. If children/pets are about, lock up chemicals and dissecting instruments in a cupboard or room.
6. Mop up any chemical spills. Do not use household cleaning products with volatiles. Copious water is best for cleaning water soluble chemicals that are spilt or use absorbent paper for chemicals that will not mix with water. Dispose of the paper in well sealed plastic bags in the 'non recyclable' waste bin emptied by the council.
7. Be aware of the specific advice given for each chemical you use, so that you know what to do if:
   a) you have a major spill,
   b) your skin comes into contact with chemicals,
   c) you swallow, inhale or get splashes in your eye. (You should be wearing goggles, so this should not happen!)
8. Be aware of flammable chemicals; never use them near a naked flame.
9. If volatile chemicals are to be heated, use a water bath or slide warmer.
10. Never heat chemicals in a test tube over a naked flame. For our purposes, when heating sodium hydroxide in a test tube it is sufficient to use a water bath.
11. Dispose of any bio-hazardous material safely. (See advice on page 4, paragraph 2.)
12. Dispose of disposable gloves "safely". (Puncture the gloves so they are unusable, wrap them in newspaper and put the package into the non recyclable domestic waste container emptied by the council.)
13. Wash your hands thoroughly when you finish work.

Page 2 to 3.
<u>Using your chemicals safely.</u>

Some notes about the chemicals commonly used in preparing for the BBKA Microscopy Certificate.

1. **Isopropyl alcohol (Propanol)** - flammable and harmful if ingested. Avoid inhaling the vapour. Used mainly as a degreasing agent and preservative in pollen slide preparation and as a dehydrating agent in anatomy slide making. Use in well ventilated area.

   <u>Spills:</u> Ventilate the area. Mop up with paper towels and dispose of in sealed plastic bags in a 'non recyclable' bin. Small spills may be allowed to evaporate providing the area is well ventilated. Wear gloves.

   <u>Skin and eye contact:</u> The odd drop on the skin will do no harm but should be allowed to evaporate. Contact with the eyes – bathe with tepid water and seek medical advice if irritation continues.

2. **Dimethyl benzene (Xylene)** -corrosive, flammable and harmful to the environment and poisonous if ingested. It may dissolve some types of plastic. Avoid inhaling the vapour. Used for clearing alcohol from specimens in anatomy slides prior to using Canada Balsam or artificial mountant. Also used for cleaning up after oil immersion microscopy.

   <u>Spills:</u> Ventilate the area. Mop up with paper towels and dispose of in thick sealed plastic bags in a 'non recyclable' bin. Small spills may be allowed to evaporate providing the area is well ventilated. Wear gloves.

   <u>Skin and eye contact:</u> Wash skin with soapy water. Contact with the eyes – bathe with tepid water and seek medical advice if irritation continues.

3. **Ethanoic acid (Acetic acid)** - in the concentrations we use it is very corrosive and irritates the respiratory tract if inhaled. Used in disease control in beekeeping***(see note on page 3)** as well as removing calcium salts in anatomy slide manufacture.

   <u>Spills:</u> Ventilate the area. If concentrated, dilute the spill with water and mop up with damp paper towels and dispose of in sealed plastic bags in a 'non recyclable' bin. Wear gloves.

   <u>Skin and eye contact:</u> Wash skin with copious amounts of water. Contact with the eyes – bathe with stream/shower of tepid water for 15 minutes and you <u>must</u> seek medical advice.

4. **Sodium Hydroxide** - very corrosive and burns skin. **Safety goggles must be worn** especially when heating this chemical. Used for macerating soft tissues when making anatomy slides.

NB. Sodium hydroxide is Caustic Soda not Washing Soda and is more hazardous.

Spills: If concentrated, dilute the spill with water and mop up with paper towels and dispose of in sealed plastic bags in a 'non recyclable' bin. Wear gloves.

Skin and eye contact: Wash skin straight away with copious amounts of water until the "soapy feeling" disappears. (The soap is produced by the Sodium hydroxide reacting with the fats and oils in your skin.) Contact with the eyes – bathe with a stream/shower of tepid water for 15 minutes and you must seek medical advice.

5. **Ethyl acetate** – flammable and has an addictive smell. Use in a well ventilated area as you may be affected by the fumes which can give you a "high". It should only be used in small quantities in your killing bottle when preparing bees for dissection.

Spills: Ventilate the area. Mop up with paper towels and dispose of in sealed plastic bags in a 'non recyclable' bin. Small spills may be allowed to evaporate providing the area is well ventilated. Wear gloves.

Skin and eye contact: The odd drop on the skin will do no harm but should be allowed to evaporate, but with larger quantities wash off with plenty of water. Contact with the eyes – bathe with tepid water and seek medical advice if irritation continues.

6. **Glycerin jelly stain (fuchsin)** - remember that this will stain clothing and skin as well as pollen! It will eventually fade and disappear from the skin!

7. **Mountant (Canada Balsam)** - contains Xylene, so treat as Xylene although it is unlikely that you will be handling it in large enough quantities to warrant taking major action. Slides can be cleaned with Xylene if application has been overenthusiastic..

***********GENERAL WARNING TO ALL BEEKEEPERS USING ACETIC ACID**********

- ***Warning! Very concentrated or glacial acetic acid is very corrosive, will if it contacts the skin cause very serious burns, is likely to cause blindness if it gets into the eyes and additionally it irritates and causes burns to the respiratory tract if inhaled. Full personal protection is essential and handling should only be carried out in a very well ventilated area.** *Glacial acetic acid is also flammable, presenting another hazard that needs appropriate control.*

Handling diseased bees.

Treat all bees as potential vectors of disease.

Diseased bees should be treated as "bio-hazardous material" and every precaution should be taken to prevent the spread of disease to other bees via you or your equipment.

Remains (of bees) should be disposed of by placing them in alcohol for 24hrs, and after draining them, they should be wrapped in plastic bags and placed in a 'non recyclable' bin. Alternatively they may be buried at a minimum depth of 1 foot or 30cms. in a part of the garden unlikely to be disturbed for some months.
Some parts may be preserved in alcohol for further treatment at a later date to process into anatomical slides.

Swill out wax dissecting dishes with propanol and allow to air dry.

Clean dissecting blocks and needles with propanol.

Any dissecting equipment that you use should be washed thoroughly in hot, soapy water – as near to boiling as possible - and dried thoroughly before storing.

Slides, mortars and pestles and other glassware should be washed in hot soapy water, rinsed and allowed to air dry. Cover slides should be disposed of in layers of damp newspaper and binned. These are sharp so take care to protect them well in the bin to prevent accidents.

Table surfaces can either be protected with newspaper (disposed of when you have finished work) or cleaned down with an antibacterial cleaner. It is safer to use a combination of both methods.

You must wear disposable gloves if you are handling diseased bees and dispose of them safely after use. (See page 1 paragraph 12)

Lab coats should be washed frequently.

These precautions should be sufficient to prevent the spread of Nosema, Acarine, Chalkbrood and Varroa via the laboratory.

Note: You should not under any circumstances be handling bees or comb infected with either AFB or EFB in the laboratory or at home. Handling bacteria safely requires training in techniques that are beyond the scope of this syllabus.

Looking after Equipment.

Your microscopes.

1. Store your microscopes in boxes or under plastic dust covers. (Dust is the worst enemy of the microscopist!)

2. Always check that your microscope boxes are securely fastened before you move them. (A microscope falling out of a box will probably be severely damaged. It may also damage you if it falls on your toe!)

3. Always use lens tissues for cleaning lenses. Solvents are not recommended as they may dissolve the cement holding the lenses in place. The exception is the oil immersion lens which should always be cleaned after use with a little xylene.

4. Always clean up spillages on the stage immediately.

5. Never place an externally powered light microscope in direct sunlight. You risk damaging your retina if sunlight is directed into the condenser or lenses.

6. Make sure you know the correct way to pick up your microscope to prevent damage to the more delicate parts

The Centrifuge.

Candidates are required to have a knowledge of how to carry out a pollen analysis of honey and this usually entails using a simple bench top centrifuge. This essential piece of equipment allows the pollen to be concentrated at the bottom of a test tube within a few minutes, rather than waiting days for it to happen under normal gravity.

They are usually straightforward pieces of kit with modern safety features that are designed to be childproof! They should have a locking device that prevents them from being opened while the rotors are still spinning (just like a spin dryer). They will only operate while a spring loaded switch is held down. Second hand centrifuges may not have these features so, if this is the case, make sure that the rotors are stationary before you open the lid. Do not use a centrifuge that does not have a lid as a tube breaking at high speed will produce flying shards of glass that can cause severe injuries.

To avoid excessive vibration it is important that the test tubes holding your samples are of equal weight. This can be done by weighing the tubes on a laboratory balance or failing that make sure that an equal amount of liquid is placed in each tube. (Providing the tubes are from the same batch and are of equal size!) If you are unlucky enough to have a breakage make sure you clean and dry the inside of the centrifuge as soon as possible.
If you are not sure how to use this machine make sure that you find someone who does and can demonstrate it to you.

Other electrical equipment.

Many microscopists make their own slide warmers and other bits of kit such as electric motor driven turntables for slide making, but make sure you have the wiring and earthing checked by a professional electrician if you have the slightest doubt about your competence in this area. This is a legal requirement if you are going to use any electrical gear in public. Microscopes with inbuilt light sources, lights, extension leads, digital projectors and especially any home-made gear must be tested and labelled on an annual basis if they are going to be used for lectures and demonstrations. Your Association should organise this 'occasion' and all electrical equipment can be taken to a central point for testing.

Safety Goggles.

Good eye protection is imperative when handling chemicals. There are many makes on the market giving variable amounts of protection. You need safety goggles that will protect you from chemical splashes - you should not need to pay more than £2 or £3. They should cover the eyes completely and have side protection. A pair of old reading spectacles or sunglasses will not offer this protection! You may need to check that you can wear your own spectacles under your safety goggles, if you need them, or alternatively you can buy prescription safety spectacles (for a price)!

Page 7.

HAZARD SYMBOLS and the new PICTOGRAMS

New pictogram. **Harmful or Irritant – may cause inflammation to the eyes, skin or other mucous membranes.**

Highly or extremely flammable with a low flash point.

New pictogram. **Compressed gases, liquids or solids.**

Explosives. Chemicals not to be heated.

Oxidising gases, liquids and solids.

Page 8.

Corrosive – chemicals that will destroy living tissue on contact.

New pictogram. **Carcinogen, mutagen, reproductive toxicity, Specific Target Organ Toxicity, Respiratory sensitisation.**

Dangerous for the aquatic environment – may present an immediate or delayed danger.

The new pictograms on the left are enclosed in a red border.
The old warning symbols have a background orange colour.

Coloured symbols can be downloaded from the second website detailed on page 9.

These pictograms or symbols should appear on bottles containing chemicals with the appropriate safety wording to act as a warning to all members of your household. Tell everyone what they mean!

(Colour reproductions of these symbols can be seen on page 46 above.)

Useful Websites.

Here is an EU based search site that advises you of the labelling and special phrases that should be used for any specific chemical:
http://echa.europa.eu/information-on-chemicals/cl-inventory-database
Under 'Search Criteria' you can enter the common name of the substance. Tick the legal disclaimer box and click 'Search'. Scroll down and under 'Search results' find the chemical you are looking for and click 'View'. A new window will open. Hold the cursor over the 'Hazard Statement Code' to see the relevant warnings. Appropriate pictograms are also displayed.

Here is a website which gives the old symbols, abbreviations and defined "phrases".
http://www.hse.gov.uk/chip/phrases.htm

Disposable gloves.

These vary in their resistance to different chemicals depending on their thickness and the material they are made from. Their resistance can be classified as:

1. **Excellent (XXX) – can be used over long periods of contact,**
2. **Good (XX) – can be used over periods of short repeated contact,**
3. **Fair (X) – can be used against chemical splashes,**
4. **Not recommended (O).**

Glove material → Chemical ↓	Natural latex (rubber)	Nitrile	Vinyl (PVC)
Acetic acid (50%)	XXX	X	XXX
Acetic acid (glacial)	XX	X	X
Propanol	XX	XXX	XXX
Xylene	O	XX	X
Sodium hydroxide	XXX	XX	XX
Ethyl acetate	O	X	O
Bio-hazards	XXX	XXX	XXX

Page 10.

<u>Risk Assessments</u>

Before beginning any practical procedure you need to identify the *hazards* that you are likely to encounter and evaluate the effect of exposure to these hazards. You should minimise the *risk* of injury or ill health to yourself and to those around you.

(A *hazard* is anything that could cause harm e.g. chemicals, fire, electricity, diseased bees, heavy objects such as microscopes or sharp dissecting instruments. *Risk* is the chance – high to low – that somebody could be harmed by these hazards. You need to evaluate how serious the harm could potentially be).

The regulations require you to give special consideration to young persons and new/ expectant mothers. When working with the public or with other beekeepers risk assessments must be recorded in writing and reviewed regularly, especially if circumstances change. Controls should be aimed at preventing accidents. Although written risk assessments are not essential if you are working on your own it is good practice to review the hazards and risks before embarking on a practical procedure.

Creating a risk assessment in five steps:

1. Identify the hazards.
2. Decide who could be harmed.
3. Evaluate the risk and assess appropriate controls.
4. Record your findings (see below) simply and clearly.
5. Review your risk assessment on a regular basis.

<u>Example of a Risk Assessment Form</u>

<u>Name</u>:
<u>Procedure:</u>
<u>Date of Assessment:</u>

<u>Review due date:</u>

1. List/ identify the hazardous chemicals/ activities/ devices/ micro-organisms that will be used.

2. Identify and assess the risks involved.

3. Describe the safety precautions/ procedures that you'll use to reduce risks.

4. Describe disposal procedures.

5. List the source(s) of safety information.

* ** *** ** *

Index

F

G

H

I

K

L

M

** ** ** **